Geology of the Yorkshire Dales

by
Peter R. Rodgers

Dalesman Books
1978

£1.5(

The Dalesman Publishing Company Ltd.,
Clapham (via Lancaster), North Yorkshire

First published 1978

ISBN : 0 85206 482 9

To David

Printed in Great Britain by
Galava Printing Co. Ltd., Hallam Road, Nelson, Lancashire

Contents

The cover photographs show:— Front: Perched boulder on Twisleton Scar, looking towards Ingleborough (Tom Parker). Back: The approach to Gordale Scar (W. R. Mitchell)

Photographs in the text are on pages 49-56.

Acknowledgement

I would like to thank the many people who have helped in the preparation of this book. My special thanks to Michael Brewster of Sheffield for producing the photographs and also to Mrs. Linda Lee.

Preface

YORKSHIRE is a county of strongly contrasting scenery and this is entirely due to the various geological events which have shaped the landscape over the years. It is hard not to find beauty in the rugged character of the Yorkshire Coast, and even the rolling scenery of the chalk wolds offers a restful natural symmetry to the human eye. The seemingly endless vales of York and Pickering are perhaps less than stimulating, but are preferable to the densely populated industrial areas in South and West Yorkshire where the scenery is dominated by tall chimneys, cooling towers, coal mines and innumerable factory buildings in unattractive 'shades' of black and grey.

It is strange, but true, that in many cases the most beautiful scenery is to be found in areas where the older rocks are exposed and this is exactly the case in Yorkshire. The areas I have already described are based upon relatively young rocks which have not yet suffered any important terrestrial upheaval. To find the oldest rocks in Yorkshire, we must move westwards away from the coast into what is undoubtedly the most beautiful area of the county, the famous Dales.

No-where else in the county can we find the same beauty, the same tranquillity, the same majesty nor the same variety. Perhaps most appreciated are the northern dales where the limestone hills and tors look down upon tree-lined valleys with their fast flowing turbulent rivers and streams. It is strange, yet the beauty of the limestone dales is enhanced by the contrast provided by the mainly barren moorland scenery which is based upon the presence of the Millstone Grit. The grit provides the seemingly endless natural battlements which often separate the dales from the moors, particularly in the south.

Of course the Yorkshire Dales are a National Park and consequently the beauty of the area should be preserved. However, the conservation of an area such as the Dales can never be assured by Act of Parliament, only the will of the people who visit the area can ensure that the conservation of the Dales in their present form will become a reality. Indeed the erosion caused by millions of visitors every year presents its own special problems in such a beautiful area. The future of the Yorkshire Dales is in our hands and each of us must play our own part.

What can the Yorkshire Dales offer you? Perhaps the scenery alone is the greatest attaction, or possibly you find an interest in the flora and fauna of the area. Certainly ramblers, climbers, naturalists and tourists are well catered for in the dales, but whatever your interest or pleasure, remember that geology is the common denominator. Everything we see, hear, touch or feel owes its very existence to the geological events which have created and shaped the landscape over the last 4,500 million years.

The story of the dales begins long before the 'ascent of man' and is full of violent and destructive events. Indeed peace has not reigned for long in the area, geologically speaking, nor will it reign a great deal longer. However, you will have time to pick up the threads of this fascinating story in 'The Geology of the Yorkshire Dales', which is about to unfold.

1. Introducing Rocks and Minerals

TODAY we are living in an enlightened world. Never before in the history of mankind have we known so much about the earth and the processes that have shaped it down the centuries. Even the process of evolution is better understood today than at any time in the past. Hardly a year passes without the publication of new ideas and information which dimly illuminates the way back into geological history.

The earth's Odyssey began thousands of millions of years ago and as the years have passed strange and violent events have taken place; events which have slowly shaped the earth into the form we have come to know today. Unfortunately we live in an ever-changing world and one day the features of the earth with which we are familiar will be swept away along the passages of time.

Before we can consider the geological events which have shaped the scenery of the Yorkshire Dales, it is necessary to consider the nature of the rocks and minerals which make up the earth's crust.

MINERALS

Minerals are the earth's building units and are defined as elements or compounds which occur naturally in the earth's crust. Elements such as gold, silver and sulphur occur in nature in an uncombined form and are therefore minerals, but other elements such as silicon and calcium only occur in a combined form with other elements. Consequently a combination of calcium (Ca), carbon (C) and oxygen (O) creates the mineral calcite, which is calcium carbonate ($CaCO_3$).

Pure minerals are scarce throughout the world and only occur in pockets, veins and fissures in rocks. The extraction of the minerals that man needs to sustain his civilisation are most easily and cheaply obtained from the relatively pure mineral veins, but many of the earth's minerals occur mainly in minute quantities which are disseminated throughout the earth's rocks.

ROCKS

Rocks are simply mixtures of minerals. Granite for example is a mixture of quartz, feldspar and mica. Each of these minerals is easily recognisable in any piece of granite. Sandstone is usually composed of grains of quartz, although on occasion the main constituent may be feldspar. Even though sandstone may be composed almost entirely of quartz, it may be coloured red due to the presence of small quantities of iron. Limestone is composed mainly of one mineral, calcium carbonate, but frequently contains a host of minor impurities which may provide colouring. Chalk, on the other hand, is also formed of calcium carbonate, or calcite as the mineral is called, but is of a much higher purity. Chalk therefore is a pure form of limestone.

Anyone who takes an interest in the countryside will be aware that there are many different types of rock and that each has its own part to play in making the scenery. Before passing on to describe the geology of the Yorkshire Dales, we must first consider the various rock types and the processes which create them.

Igneous Rocks

Igneous rocks are subdivided into two groups, intrusive and extrusive, but rocks of both types originate within the earth, only their cooling characteristics are different. The name Igneous comes from the Latin 'ignis' meaning fire, so the character of these rocks is fairly obvious.

Extrusive rocks are those formed by volcanic action. The molten rock which is called 'magma' wells up from within the earth and is expelled by a volcano, frequently in a violent manner. On coming into contact with the earth's atmosphere the molten rock cools rapidly and forms a dark-coloured, fine grained rock. Although rocks of this type vary widely in chemical composition and contain a range of minerals, their rapid cooling prevents the minerals growing to visible proportions. The result is that the rock appears to be composed of one mineral when viewed by the naked eye. Microscopic examination however, reveals all the minerals of which the rock is composed.

Intrusive rocks originate from magma within the earth in exactly the same way as the extrusive variety, but these are not extruded by a volcano. If magma rises from within the earth towards the surface but becomes trapped within the rocks of the earth's crust, the magma will cool exceedingly slowly. The result is that the minerals in the slowly cooling magma have the time to develop large crystals. Intrusive rocks may have the same variety of chemical composition as the extrusive or volcanic rocks I have already described, but due to their slow rate of cooling they have a light colouration and the individual constitu-

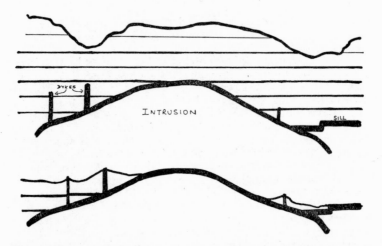

The formation of igneous intrusions. Top: The molten magma has intruded its way into the country rock and has formed a number of dykes and a sill. Bottom: Erosion has largely removed the country rocks, exposing the intrusion to the atmosphere. The dykes are harder than the country rock (limestone) and stand proud from the earth's surface as small ridges.

ents of which the rock is composed are clearly visible to the naked eye.

Granite and gabbro are perhaps the best known representatives of the intrusive range of igneous rocks, while the extrusive rocks are predominantly represented by basalt and andesite. In practice there are many different types of rock within the two groups and some degree of overlap occurs. A typical example is presented by the igneous rock structures called sills and dykes. These are intrusive in character, but frequently, due to their small proportions, they cool rapidly enough to be similar in appearance to volcanic rocks.

Sedimentary Rocks

Sedimentary rocks owe their character to the earth's erosive forces — forces which we experience everyday of our lives, but whose destructive nature we often fail to appreciate. The destructive nature of wind and water is, however, fairly obvious. How often during recent times have we seen houses perched precariously on the edge of cliffs soon to topple over on to the beach? Such incidents provide more than adequate evidence of how rocks, in this case cliffs, may be destroyed by the abrasive action of the sea. The wind also plays its part in the destruction of rocks, as the particles it carries about erode the exposed rock

9

surfaces. Rainwater also adds its destructive action and we must not forget the alternation of hot and cold provided by summer sun and winter frost.

It is the destruction of mountains and continents by the forces of erosion which makes the formation of sedimentary rocks a reality. The rock debris eroded from mountains and continents is washed by rain, streams and subsequently rivers towards the sea. On reaching the sea the debris sinks to the sea floor and remains there, slowly accumulating. The coarser river deposits are deposited in or close to the river mouth, the successively finer particles of debris being washed far out into the ocean. As the years pass away, the slowly accumulating sediments build up in depth on the ocean floor. Eventually compression, either by sheer weight of accumulated debris or due to the onset of a mountain building orogeny, transforms the debris into a sedimentary rock. However, no actual modification of the character of the sediment takes place. The whole mass of sediment simply becomes bonded together. Sandstones, shales, siltstones, gritstones, conglomerates, breccias and some mudstones may be formed in this manner. The rock fragments are usually held together in a matrix of quartz, calcite or iron oxide. Sedimentary rocks formed from eroded debris are called clastic sediments.

The sedimentary rocks already described can be sub-divided by the size of the particles of which they are composed. Conglomerates and breccias are composed of large debris which in the case of the conglomerates are usually nice round pebbles. Breccias are formed from angular fragments. Gritstones are sandstones in which the particles are angular or sub angular. Sandstones vary in grain size from 2 mm which is considered very coarse to 1/16 mm which is extremely fine. A rock with particles less than 1/16 mm is termed a siltstone and when particle size gets below 1/256 mm the rock has reached clay grade. Mudstones and shales both consist of clay grade particles, but shale exhibits bedding planes while mudstones do not. The term marl is often used to describe a calcareous (lime rich) mudstone.

Sedimentary rocks may be formed by different processes and the resulting rocks are called organic and inorganic sediments. Organic sedimentary rocks are formed from the remains of creatures which thrived in the seas many millions of years ago. Life has taken many millions of years to evolve to its present forms, but it was only around 600 million years ago that creatures, all of which lived in the sea, began to develop hard shells or skeletons. This was done by secreting lime from sea water. As the years passed away creatures such as corals were to come to the fore and in the sparkling clear waters large colonies thrived. When the corals died their skeletons remained and finally, due to mountain building and continental drift, the remains of the

10

corals were brought to the earth's surface in the form of lime-stone. Of course all creatures with hard parts would, on dying, contribute their remains to the growing calcareous sediment on the sea floor. Limestone and chalk are the predominant rocks formed in this way, although chalk is formed from the remains of cocoliths and foraminifera. These were tiny creatures which lived in colonies like corals. The chalk scenery of eastern England is nothing more than a giant epitaph to these minute sea creatures which lived around 100 million years ago.

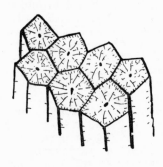

**Corals from the Carboniferous. Left: Lithostrotion (Rugosa).
Right: Zaphrentis (Rugosa).**

Organic sediments are formed from the remains of past life, but inorganic sediments are formed as a result of changes in the climatic conditions existing at the time. When the climate is arid, lakes and seas begin to dry up due to evaporation. As the water level falls, small areas of water may become separated from the main mass. If the hot conditions persist, the water will continue to evaporate until the minerals dissolved within the liquid become highly concentrated. Eventually the liquid becomes supersaturated with the minerals and they are precipitated to the bed of the lake. From time to time the sea may reassert itself and replenish the lake, only for it to evaporate once more. The repeated precipitation of minerals can result in the forma-tion of large quantities of evaporites which may take the form of limestone, gypsum, halite and anhydrite. The economically important salt deposits of Cheshire were formed in this way.

All sedimentary rocks tend to exhibit a banded structure known as bedding; each bedding plane corresponding to a change in the environment at the time the sediments were being laid down. When the bedding planes of an exposed rock lie in one plane and another rock by which it is overlain lie at a totally different angle, then an unconformity is said to exist. This means that there is a break in the rock succession. In other words the lower rock had been subjected to erosion before the rock above it had been laid, so the rocks were not the result of continuous deposition. A fine example of this can be seen at Thornton Force near Ingleton in the Yorkshire Dales. However another factor which is of great importance is that sedimentary rocks are the only type of rocks which yield fossils.

Fossils

Fossils are the aid to understanding the process of evolution and to the dating of rocks. Although there are one or two exceptions, fossils are confined to sedimentary rocks. A fossil is defined as evidence of past life which has been preserved either in its original form or by replacement by rock or mineral. Many sedimentary rocks contain fossils in great quantities, but the number of fossils when compared with the vast numbers of creatures which have lived on the earth is dismally small. Fossilisation, therefore, has been a rare occurrence.

Fossils of creatures which lived in the seas and oceans of the world are most common. Living and dying beneath the waves, their skeletons were easily entombed in the sediments being accumulated on the ocean floor. Creatures which lived on land were not in the right environment for their remains to become fossils, consequently the fossils of amphibians, dinosaurs and mammals are rarer finds. The original structure of the creature is often destroyed during the rock forming process and is replaced by a mineral. The most common mineral to take part in the replacement is calcite. However, rocks may also form fossils, good examples being offered by limestone and chalk.

Fossils are only of value to palaeontologists when they can be related to the horizon in the rock in which they were preserved. An understanding of the evolutionary pathway can then be used to date the fossil and also the rock in which it was found. Early forms of life did not have the hard skeleton which could readily become fossilised, consequently fossils of the earliest life forms are very scarce. The development of skeletons became more common around 600 million years ago and fossils are therefore primarily found in sedimentary rocks which were formed during the last 600 million years.

Metamorphic Rocks

Metamorphic rocks are extremely varied in type and are formed as a result of the earth's most violent and tempestuous processes. These rocks begin life as sedimentary or igneous rocks and are subjected to great heat and pressure which effectively primarily found in sedimentary rocks which were formed during times of crustal stress when mountain chains are being created. Any rock, close to the heart of an orogeny is bound to suffer extensive metamorphism and will undergo modification into a different rock. The minerals of which the rock is composed have the opportunity under great pressure to recrystallize, hence the formation of a new rock. Varying amounts of heat and pressure can produce varying degrees of metamorphism in the rock. Inevitably this type of metamorphism is tied up with plate tectonics. When metamorphism takes place on a grand scale, resulting in the modification of large areas of rock, the rocks are said to have undergone regional metamorphism.

Metamorphism may also take place on a small scale when heat is the factor which brings about the modification of an existing rock. Contact metamorphism takes place on a much smaller scale and is directly due to the effects of igneous activity. When molten rock, or magma as geologists call it, begins to eat its way into the earth's crust, it comes into contact with existing rocks. The heat from the magma is sufficient to modify the character of any rock with which it comes into contact.

Of course the size of the igneous body will greatly influence the degree of metamorphism that the adjacent rocks undergo. In the case of a dyke which measures little more than a foot in width, the metamorphosed area of the adjacent rock may be minimal. However, when the igneous bodies are large, then the metamorphic aureol may be very extensive. Let us consider Cornwall as an example. Here the heat from the granite of Lands End, Carnmenellis and Bodmin has greatly altered the rocks into which the granite magma was intruded, but that was many millions of years ago.

Glacial Rocks

The great ice age of the Pleistocene began approximately two million years ago. It was heralded by a cooling of the climate in the northern hemisphere and before long snow was falling over many areas of Europe. As the years passed, the summer sun failed to melt the snow from the preceding winter and large quantities of snow were built up on high ground in mountainous areas. Eventually the snow was compacted into glacier ice and gravity began to draw the ice off the mountains. The cold icy fingers of the glaciers began to spread out across the countryside. Once on the plains, glaciers came together to form ice sheets which progressed foot by foot inexorably further south with every passing winter. Eventually the ice reached a latitude beyond which the temperature was too high for it to exist and on reaching this point, the ice melted. Warmer interglacial periods took place when the ice retreated back to the mountains from whence it had come, but each time the climate would reverse and once again the ice would reach out towards the south.

The ice covered Britain on no fewer than three occasions, but it never moved south of a line from the River Severn to the Thames. As the ice moved over the countryside so it transformed the scenery. The weight of a hundred feet of ice scraping over the rocks of the earth's crust was bound to cause considerable havoc. Rock fragments and boulders were broken off the mountains and hillsides, and were carried along beneath the ice. As the ice travelled considerable distances, it is hardly surprising that great quantities of rock debris were accumulated under the ice. When the ice melted, the debris was left as a sign of its passing. Today this glacial debris has taken the form of a thick clay called boulder clay and within the clay are many rock fragments which were carried to their present position by the ice sheets long ago. The result of this process is that rocks of Scottish and Scandinavian origin may now be found in parts of Britain many miles from where they originated.

When the ice melted, rivers flowed out from beneath the ice sheets and carried the rock debris over the countryside. Accumulations of glacial river gravels are called fluvioglacial deposits and are worked commercially in many part of Britain. Glacial deposits are rarely found on high ground, for being soft, they are easily eroded by the elements. Boulder clay may be observed overlying the upper beds of hard rock in a quarry. The same material may underlie the soil on agricultural land, and the pebbles and rock fragments it contains may be found in fields, much to the frustration of many a farmer.

PLATE TECTONICS

Although by no means a new concept, the processes of plate tectonics and continental drift have received considerable publicity during the last few years. There can be very little doubt that the continents of the earth do move about, although only exceedingly slowly. The general pattern is for the continents to come together forming more or less one land mass, and when this has been achieved, they separate ultimately to form yet another great super continent.

The continents are situated on plates which transport them. On occasions when the continents are coming together, the plates collide which produces the great upheaval of rocks necessary to form a mountain chain. It would appear, therefore, that the great mountain building periods of the earth's history were the direct result of continental drift. These processes are still active today. At this moment Africa is moving northwards into Europe. The Mediterranean Sea, the cradle of our civilisation, is dying.

THE GEOLOGICAL TIME SCALE

Eras	Period	Epoch	Millions of years since start of period
CAINOZOIC	Quaternary	Recent (Holocene) Pleistocene	2
CAINOZOIC	Tertiary	Pliocene	12
CAINOZOIC	Tertiary	Miocene	25
CAINOZOIC	Tertiary	Oligocene	40
CAINOZOIC	Tertiary	Eocene	60
CAINOZOIC	Tertiary	Paleocene	65
MESOZOIC	Cretaceous		135
MESOZOIC	Jurassic		180
MESOZOIC	Triassic		225
PALEOZOIC	Permian		270
PALEOZOIC	Carboniferous	Coal Measures Times Millstone Grit Times Carboniferous Limestone Times	350
PALEOZOIC	Devonian (Old Red Sandstone)		400
PALEOZOIC	Silurian		440
PALEOZOIC	Ordovician		500
PALEOZOIC	Cambrian		600
	Pre Cambrian Origin of the Earth		Undefined Over 4,500 million years ago

One day, perhaps millions of years hence, this beautiful sea will have been replaced by a great mountain chain. But however great the mountains, one day they in their turn will be swept away on the whim of our ever-changing planet.

THE GEOLOGICAL TIME SCALE
(see page 17)

The story of the earth's history is so long that it has been deemed necessary to divide the story into smaller sections, hence the development of the Geological Time Scale. Reference to this scale is repeatedly made throughout the book. Some of the names of the periods were derived from the areas where the rocks were first studied, for example, Devonian—Devon, Cambrian—Wales; others came from the rocks formed during the period, for example, Cretaceous—chalk.

Key to Geological Map of the Yorkshire Dales

Igneous rocks of Carboniferous, Permian and Tertiary ages in the Carboniferous Limestone Series.

Rocks of Triassic age.

Rocks of Permian age.

Rocks belonging to the Coal Measures (Westphalian).

Rocks belonging to the Millstone Grit Series (Namurian).

Rocks belonging to the Carboniferous Limestone Series (Dinantian).

Rocks belonging to the Lower Palaeozoic.

Locations
A Askrigg
C Clitheroe
H Harrogate
I Ingleton
M Middleton in Teesdale
R Richmond
S Sedbergh
SK Skipton

16

Geological Map of the Yorkshire Dales.

2. Setting the Scene

THE GEOLOGY of the Yorkshire Dales is inextricable from that of the rest of the British Isles and forms but a very small part of a very complicated picture. The county boundaries created by man are meaningless when viewed in terms of the geology of the countryside and although the Yorkshire Dales form the central theme of the book, it is inevitable that from time to time the story will assume a broader basis.

The rocks which outcrop in the Yorkshire Dales belong almost entirely to the Carboniferous Period of geological history which was in full swing some 300 million years ago. However, the events of the period were almost a foregone conclusion for the scene had been set many millions of years before. 4,500 million years ago the earth was in existence as a glowing ball of gas at extremely high temperatures. For millions of years the gas cooled before its character was changed and the gas was transformed into a semi-molten state reminiscent of molten rock. The surface of the earth cooled and the first rocks began to form. The earth's crust thickened as the temperature continued to fall and on reaching the critical temperature, water was formed and it rained for thousands of years.

We now move into a period of uncertainty because in no instance have the rocks formed during the early days of the earth been preserved. Subsequent earth movements have first modified them out of all recognition and then erosion has removed them from the scene. The oldest rocks exposed in Britain which form the Island of Lewis in the Hebrides are dated as 2,800 million years old and are but a shattered remnant of an old continent broken up many millions of years ago. The truth is that the earth's crustal movements have so disguised the ancient rocks that the pieces of the geological jigsaw puzzle are impossible to fit together. It is only in younger rocks which have not been subjected to major crustal deformation that we can piece together a story. Consequently the happenings on earth before 600 million years ago are speculation, nothing more. Events which have taken place during the last 600 million years are more accurately tabulated.

We can be certain that the continents of the earth were adrift long before 600 million years ago and that mountains and oceans have been created and destroyed on many occasions before. Mountains unclimbed and oceans unnamed have littered the face of the earth countless times in the past and few have left any sign of their former presence. Although the rock structure of the British Isles has come from a number of different sources, we can say that 500 million years ago Britain was well to the south of the equator and during this 500 million years the progress has been northwards, although it has been subjected to a number of delays en route.

The story of life on this planet also dates back into the distant geological past and here too information is sparse. It was only when the life forms began to develop hard skeletons which could be easily fossilised that fossils became plentiful. Consequently, our knowledge of life itself is not well tabulated beyond 600 million years ago. Life was becoming abundant in the seas 600 million years ago, but it would be another 200 million years before the first amphibians would stagger out of the water on to dry land.

If we could turn back the clock 500 million years to the geological period geologists have called the Cambrian, Britain would not exist as we know it today. Indeed at this time much of Britain was under water. A land mass of continental proportions probably lay away to the north, and the erosion of this land mass was providing sediments which were being washed out onto the sea floor which was slowly subsiding. As the years passed, pressure was exerted upon the area around the sea trough and slowly it began to narrow. The buckling of the sea floor marked the beginning of the Caledonian Orogeny. The word orogeny is used to describe all the events which take place when the continents collide and mountains are being created.

The narrowing of the sea trough covering Britain began at the end of the Cambrian Period and the subsequent Ordovician Period was typified by intense volcanic activity. The volcanic rocks of the Lake District and Wales belong to this period of vulcanism. Naturally the sea floor was not of uniform depth, the buckling created by the narrowing process serving to create shallows and deep water basins. Indeed in some areas the sea floor may have risen above sea level for a time.

The Silurian Period began around 420 million years ago and was similar in character to the Ordovician, although the volcanic activity was considerably reduced. The sea trough continued to narrow and as the deep water basins became infilled with sediments, so shallow water deposits were the principal rocks laid down. No major land mass existed near the British Isles and in clear water conditions limestone was formed for a time. The latter

The geography of Britain during Silurian times.

The geography of Britain during Devonian times.

part of the Silurian Period was marked by the onset of the main earth movements of the Caledonian Orogeny.

Slowly but surely the intense pressure on the rocks of the earth's crust caused them to move. Indeed, although by our standard it was exceedingly slow, the tortured rocks began to bend, buckle and fold. The Caledonian Mountains rose out of the sea trough and although their rise to stardom was exceedingly slow, they became a considerable scar on the landscape. Indeed these mountains have been likened to the Himalayas for size, although this can be no more than an educated guess, as man would not appear for many hundreds of millions of years by which time the mountains would have been ground into dust. However, this magnificent mountain chain was to provide the key to future events and would lead to the creation of the rocks which are exposed in the Pennines today.

The beginning of the Devonian Period coincides with the time that the Caledonian Mountains reached their full height. The mountains covered most of Northern Ireland, Scotland and Scandinavia. Northern and Central England were part of the foothills of the great mountains. Parts of Southern England were under water at this time. The uplifting of the mountains continued for some time and was accompanied by considerable volcanic activity in Scotland and the North of England. However, erosion was also taking its toll and the desert lowlands between the mountainous tracts were becoming choked with debris as the mountains were abraded.

During the 50 million years of the Devonian Period the mountains of England and Wales became worn down almost to sea level, but in the far north the Caledonian Continent was still an impressive structure towering over lowland Britain. Eventually the desert countryside of much of England and Wales subsided below sea level, but by this time the Carboniferous Period of Geological History had dawned. The time is 350 million years ago. Thanks to 'continental drift', by the Devonian Period, Britain was south of the equator on the desert belt, hence the hot arid conditions, but by the beginning of the Carboniferous Period, the country was very close to the equator and still moving north.

Direct evidence of the Pre-Carboniferous geology of the Yorkshire Dales is difficult to find today, but in one or two places small inliers of Pre-Carboniferous rocks can be found. Silurian rocks are represented in the Dales near Horton-in-Ribblesdale and near the county boundary on the Howgill and Middleton Fells. Rocks of this age include shales, slates and grits. Lower Palaeozoic Rocks are also exposed near Cautley where they include a volcanic formation amongst mudstones and shales. The sequence here is reasonably complete and is therefore superior to the other inliers around Sedbergh and Craven. Unfortunately the age of these rocks is still somewhat in doubt. There has been considerable disagreement over the years as to whether the oldest rocks exposed in Yorkshire are of Ordovician or Cambrian Age.

Rocks of Devonian Age are not believed to be exposed in the Dales although basement beds, mainly conglomerates, at present attributed to the Lower Carboniferous may well be of Upper Old Red Sandstone Age. While few exposures of rocks other than Carboniferous Age are exposed within the Yorkshire Dales, it is thought that a basement of older rocks, possibly of the Skiddaw Slate Group, may traverse much of the county at a depth below the Carboniferous cover.

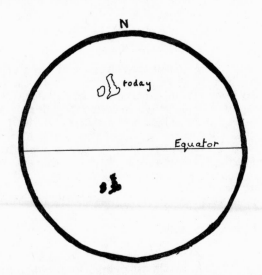

The position of the British Isles on the earth's surface during Devonian times was well south of the equator.

3. Carboniferous Limestone Times

SOME 350 million years ago as the Carboniferous Period of Geological History dawned, the British Isles was situated very close to the equator. Much of the country was out of the water but most of southern Scotland, England and Wales would be under water for much of the next 40 million years. Only the north of Scotland and the Welsh Mountains were to avoid the great flood to come.

The remains of the great Caledonian Continent was by now a relatively low lying tract of land although it still assumed continental proportions. It appeared far removed from the mountainous structure it had been during late Silurian and early Devonian times. The arid desert landscape of the Caledonian Continent was still largely devoid of vegetation, although plant life was soon to become increasingly prolific around the edges of a new sea. The life under the sea had developed to a sophisticated level by now and the first amphibians had already staggered out of water onto dry land.

Although relatively stable for a long period, the earth's crust in the Yorkshire area began to subside exceedingly slowly. The lowlands of the Old Red Sandstone Continent continued to subside for quite a long period until as the land became lower, so the sea waters began to encroach upon the desert scene. It would be nice, indeed it would be almost poetic, to say that the deserts were drowned beneath the waves, but in geological terms events move very very slowly and it would be wrong to imply that the sea rushed in.

Eventually much of southern Britain was under the waters of a new sea, the Carboniferous Sea. There were however quite large areas that the waters did not cover. Subsidence of the sea floor was anything but uniform and a number of stable, block-like areas existed which were to resist the advance of the waters for a long period of time. The main areas of stability in the Yorkshire area, although others existed, are the Alston and Askrigg Blocks. The Alston Block, except for its southern extremity, is of course outside Yorkshire. It is envisaged that the two blocks were partially separated by the Stainmore Trough. Although deep, this

The geography of Britain during Carboniferous Limestone times.

The relationship of the Alston and Askrigg Blocks during Carboniferous Limestone times.

trough was not to assume a basinal character and contain basinal sediments, it would normally only represent a more complete succession of the sediments which were deposited around and occasionally on the blocks. To the north, the Southern Uplands of Scotland were probably never entirely covered by the sea, while away to the south the Welsh Mountains formed the highest part of a land mass which today has been called St. Georges Land. This was a land barrier which at one time extended from the Wicklow Mountains of Ireland through Wales and England into Belgium.

The scene is now set, but before going on to consider the geological happenings of the next 20 million years, it is necessary to consider briefly the terms geologists use to describe the various parts of the Carboniferous system. The Lower Carboniferous or

Dinantian is described as two series, the Tournaisian and the Visean. Six stages of cyclic sedimentation are recognised in the Lower Carboniferous, these being numbered 1 to 6. Names have not yet been allotted to the stages.

Stratigraphical Divisions of the Lower Carboniferous
(Dinantian)

Series	Stage	
Visean	6	
	5	Carboniferous
	4	Limestone
	3	Times
	2	
Tournaisian	1	

In the first cycle, number 1 stage belongs to the Tournaisian, while the remaining 5 stages belong to the Visean. Stages 1, 2, 3 and 4 are considered to be major cycles. Stages 5 and 6 are made up of several lesser or minor cycles. The Upper Carboniferous is also sub-divided, but these terms will be considered in the appropriate chapter.

Invasion or transgression by the Carboniferous Sea resulted in sediments being laid down in the various basins. However, different types of sediment were being laid down in different basins of the sea. In the north, in central Scotland, eroded debris from the Caledonian Continent was building up beneath the surface of the sea. Later these sediments would be transformed into the shales, siltstones, cementstones and sandstones which are exposed in the Midland Valley of Scotland today. The Carboniferous Limestone, which was the characteristic rock formed during the Lower Carboniferous is only important further south. In Scotland, the limestone occurs only in thin bands and is only a minor part of the succession.

Progressing southwards, the waters of the Carboniferous Sea became clearer. This was due to the increasing distance from the northern continent. Most of the eroded debris from the north was deposited close to the continent, only the lighter particles of debris being washed southwards. In addition, the Southern Uplands and the stable Alston and Askrigg Blocks would also serve as a barrier restricting the passage of sediment to the south. Of course some sediment would be derived from these stable areas but, being low lying, this would only be important locally if at all.

TOURNAISIAN

In Northumberland the Lower Carboniferous is first represented by shales and sandstones as well as cementstones, otherwise called Magnesian Limestone. Unfortunately the base of this, the first cycle of Carboniferous sedimentation, is nowhere exposed. The oldest rocks formed during the cycle of marine invasion, or transgression, which are exposed in Yorkshire, are to be found in what geologists call the Craven Lowlands. The Craven Lowlands is the name given to describe the area of Yorkshire which formed part of the Bowland Basin. This basin was the main area of deposition south of the stable blocks and its centre was situated somewhere to the south and west of Yorkshire. The floor of this basin, like that of the Stainmore Trough, was actively subsiding at this time.

These Tournaisian rocks also probably occur in the Stainmore Trough, but this is unsubstantiated. However, the trough is said to contain the same sequence as is today exposed in Cumbria in Ravenstonedale; a fact which may be used for correlation. The oldest rocks known in Yorkshire which represent this first marine transgression are shales and dark bituminous limestones. As I have already indicated, rocks of the first transgression by the sea are not well exposed in Yorkshire, but a depth of over 300 metres has ben attributed to these rocks in the Broughton area.

The first regression by the sea was to produce algal limestones in the basinal area and by the formation of the Caldbeck Limestones, which include algal growth and thin shale bands in the North.

VISEAN

The sea floor continued to subside unevenly during the second cycle and the sea only managed to make inroads into the basins such as the Bowland Basin, and the Stainmore Trough in Yorkshire. It would be some considerable time before the sea would begin to lap around the edges of the stable Alston and Askrigg Blocks.

The transgressive part of this cycle produced almost 200 metres of dark limestones which were largely made up of broken fragments of organic skeletal material derived from creatures living in the water at the time. Shales are interbedded amongst the limestones, but are not present in any thickness. Indeed, shales are more prominent in the Skipton area and it is considered that this area may have been more or less immune to subsidence at this time. In Ravenstonedale, which is believed to mirror the geology of the Stainmore Trough, the readvance of the sea at the beginning of the second cycle resulted in the formation of nearly 120 metres of dark limestones.

As the unstable sea floor again asserted itself, the second cycle of marine transgression came to an end. The regression of the sea is confirmed by the character of the sediments which today overlie the dark limestones. In the Stainmore Trough, as represented by Ravenstonedale, the regression of the Carboniferous Sea is outlined by the presence of oolitic limestones, sandstones and pebble beds; all shallow water deposits. Calcite mudstones and reef knolls provide evidence of marine retreat in the northern part of the Bowland Basin (Craven Lowlands).

By all accounts this was a major regression of the sea but the third cycle, the second in the Visean, was to see the waters make considerable inroads on to the resistant Askrigg Block. The third advance by the sea was marked by the formation of 230 metres of dark limestone in Ravenstonedale. At Salt Hill near

Crinoids were a relative of the sea lily of today.

Clitheroe, in the Craven Lowlands, the first rocks created by this marine advance are highly fossiliferous, containing vast numbers of crinoids, which have been described as fossil sea lilies; but further south towards the centre of the basin, the limestones give way to shales. The crinoids are common fossils in the Carboniferous Limestone wherever it occurs. These creatures were similar to the sea lilies of today having a root and a stem made up of many tiny columnals, surmounted by a flower. Certain varieties were capable of uprooting themselves and floating about, dropping anchor, as it were, and re-rooting themselves at their own will. Fossil crinoids in the Carboniferous Limestone are usually represented by many columnals; remains of the flowers are not normally found.

The Carboniferous Sea only managed to creep onto the southern edge of the Askrigg Block towards the end of this transgressive phase. Consequently the sediments deposited in this area have achieved little thickness and are limestone which contain many fossils. It is this fossil content which has enabled palaeontologists to conclude that the limestone was formed late in the third transgression. Fossils found on the southern edge of the Askrigg Block, when related to the fossils in the first or basement beds of the cycle, do not correspond. However, they are similar to fossils in the upper beds of the transgression elsewhere in Yorkshire. The principal fossils used for dating the Lower Carboniferous rocks in Yorkshire are goniatites which are a type of lamellibranch. In other areas where Lower Carboniferous rocks are exposed, other fossils including corals and brachiopods are used.

The subsequent regression by the sea produced sandstones and dolomitic limestones rich in gastropods on the southern edge of the Askrigg Block, but in the Craven Lowlands this regression is marked by a cementstone bed within beds of shale.

The end of this, the third invasion by the sea, was by no means the last chapter in this fascinating yet complex story, for indeed the sea would return and recede many more times before it would be gone forever. On occasions the story of the Yorkshire area can be traced through the rocks and easily understood, but more often the story is camouflaged by the numerous geological events which have taken place since the Carboniferous Sea invaded the British scene.

The fourth major invasion by the sea resulted in the formation of limestones in Ravenstonedale, and by inference in the Stainmore Trough, but there is evidence that the sea receded partially, at least four times during the cycle. The same limestones occur near Ingleton and floor the dales on the south side of the Askrigg Block. Fossils, particularly crinoids, are abundant in limestones lower in the sequence, but in many cases the higher beds are relatively barren. In the Craven Lowlands the transgressive phase is marked by a dark grey limestone containing chert bands, but further south, towards the centre of the Bowland Basin the entire sequence is represented by shales. It is important to note that limestones of the third, fourth and fifth cycles in the Settle area, are collectively known as the Great Scar Limestone.

The regressive, or retreat phase, of the fourth cycle is represented by a fine grained limestone which is present over much of the southern part of the area. The Pennine Fault Block is made up of the Askrigg and Alston Blocks which were to some extent separated by the Stainmore Trough and continued to present a major barrier to the sea. In fact the Askrigg Block was only submerged towards the end of the fourth phase and the Alston Block would survive until the next major invasion by the sea.

Yet again the sea returned to begin the fifth cycle of sedimentation and limestones were again formed. The lower limestones of this phase are only to be seen near Kirkby Stephen and Sedbergh. Thick bedded, fine grained limestones represent this cycle near Ingleton, Ravenstonedale and on the Alston Block. These limestones are thickly bedded, each bedding plane representing a marginal change in environment created by a regression or transgression by the sea. There were in fact no fewer than nine minor regressions and transgressions by the sea, each resulting in the deposition of limestone with thin beds of shale. The Alston Block and the southern edge of the Askrigg Block were not covered at the beginning of the fifth transgression and consequently the lowest beds of the sequence are missing.

It is generally considered that the limestones were being formed in shallow water and that the limestone surfaces in some cases may have emerged slightly from the sea and been subjected to erosion. The shallow character of the sea at this time is also testified to by the presence of a number of thin coal seams amongst the limestone near Ingleton. The plant life necessary to create coal could only have existed in the shallowest of water or on virtually dry land. It is also interesting to consider that coal forest-type situations could be created at this time during the Lower Carboniferous.

The limestones formed during this cycle play an important part in the scenery of the southern Dales of Yorkshire today, forming as they do, many of the limestone scars which greatly beautify the area. Not for nothing has limestone of Carboniferous Age been called 'Mountain Limestone'.

However, in the Carboniferous Period 300 million years ago, the mountains that mattered lay several hundred miles away to the North. Uplift of the continent which lay across northern Scotland and Europe combined with a wet monsoon-like climate, served to increase the rate of erosion of the northern hills, and this would affect the environment in the Carboniferous Sea. A river, or number of rivers, draining the Continent in the North, became faster flowing due to uplift of the landscape, and were transporting fine river debris into the sea. The lighter particles were carried far to the south and gave rise to the formation of rocks substantially different from the limestones. Indeed, sandstones and shales occur between limestone beds on the Alston Block and one variety called the Thorney Force Sandstone can be found in Wensleydale. These sandstones and shales represent periods of turbidity when the Carboniferous Sea became cloudy on account of the presence of fine eroded debris which had originated on the Northern Continent. These rocks are known to geologists as Yoredale Rocks. The limestone beds between the sandstones and shales represent periods of time when the water was clear and free from the elastic sedimentary particles.

Another interesting feature of the rocks formed during the fifth cycle of deposition is found in the reef limestones which occur immediately to the south of the Middle Craven Fault on the south of the Askrigg Block. These knoll reefs are still the object of considerable controversy in geological circles today. It is considered that the reef belt extends to both east and west, adjacent to the faults, although this can only be a surmise as the reef limestones are covered by younger rocks.

Several theories have been put forward to account for the formation of the reefs, but it is now considered that the reefs began during the regression of the sea at the end of the fourth cycle of marine transgression and were simply a reef apron at the edge of the Askrigg Block to the north. However, other knoll reefs are also said to have grown individually on the sea floor and may have developed to a height of 175 metres.

In the Craven Lowlands the sediments laid down during the fifth cycle are different from the limestones of the Askrigg Block. In fact limestone, although present, does not have the importance that it holds in the northern Dales. Shales and mudstones are the principal rocks, the limestones being subsidiary. A number of breccia beds occur in the limestone and have been linked with the reefs. It is envisaged that the breccias were formed from scree material from the reefs.

Although the sixth cycle of deposition is considered to be part of the Lower Carboniferous, it is in fact almost entirely devoted to the Yoredale Series of rocks. The days of a shallow, sparkling and clear Carboniferous Sea were almost over and it was soon to be gone at least from the southern area for ever. The sixth cycle is characterised by an often repeated sequence of rocks which signify the southerly movement of a river delta and subsidence of the sea floor; the cyclic sequence being limestone, shale, sandstone and coal.

The sandstones and shale of this cycle become thinner as one proceeds southwards and the limestone comes once again into its own. Naturally the deltaic debris from the north tended to die out southwards as the distance from the northern shore was increased. However, it is also acknowledged that there was significant movement of the Middle Craven Fault and the southern edge of the Askrigg Block was uplifted providing a barrier to the southerly passage of sediment from the north. It is generally considered that the Yoredale Rocks represent the advancing and retreating front of a river delta, the shales, sandstones and silt-stones being debris from various areas of the river delta. The limestones continued to be formed in areas where either sea floor subsidence had taken place or marine transgression had ensued. The sedimentation south of the Craven Fault is still represented by shales and mudstones which belong to the sequence known as the Lower Bowland Shales. Age correlation between rocks of the Lower Bowland Shales in the basinal area south of the Craven Fault and the rocks of the Yoredale series to the north have been made, but during the sixth phase of sedimentation the uplift of the southern edge of the Askrigg Block prevented, or at least minimised, any movement between the two areas.

By this time the life of the Carboniferous Sea was coming to an end. The clear sparkling waters and the limestone they created would survive in other areas for some time to come, but in the basinal area south of the Pennine Fault Block, the sea in its Lower Carboniferous form would soon be gone forever.

The mountains and rivers of the continent in the north would become increasingly important in the years ahead and, in part at least, the rocks still to be created would have very important uses for man. However, with the geological clock fixed at around 300 million years ago, man would not make his appearance for some considerable time.

We have imagined the cool clear waters of the Carboniferous Sea, waters in which limestone was created. I have described how the sea floor was subsiding in some areas and stable or being uplifted in others, and how the sea suffered a change in environment as the fast-flowing rivers of the northern continent carried eroded rock debris southwards hundreds of miles from their source. Finally, I have described the Carboniferous Sea becoming increasingly cloudy as the river delta pushed its way into the area. The limestone forming conditions were gone, their place was taken by shale, sandstone and mudstone. In such a changing world one might well ponder over what lies ahead.

4. Millstone Grit Times

THE Lower Carboniferous period came to a close at the end of the sixth phase of sedimentation in what geologists describe as the Visean series. The Upper Carboniferous begins with the Namurian Series which geologists have divided into seven different stages. Each stage is distinguished by the appearance of a particular goniatite amongst the mudstones.

First however the scene must be set. The end of the Lower Carboniferous era and the beginning of the Upper Carboniferous would not have been obvious to any observer, casual or otherwise, who happened to be around, 300 million years ago. Indeed, the junction is marked not by volcanic, intrusive or any tectonic action for in fact it occurs at the point in time that the small goniatite Cravenoceras Leion first appeared on the scene. The landscape of the Yorkshire Dales at this time was still that of a shallow sea, but the river delta from the north was soon to invade this part of England and the shallow sea would become choked with mud and sand banks.

The principal rocks formed during Millstone Grit times (Namurian) were sandstones which include the coarse varieties known locally as gritstones. The sandstones are interbedded with mudstones and siltstones. Mudstones of marine and non-marine origin occur and their differentiation is due, not only to fossil content, but also to colour; the marine mudstones being black, their non-marine counterparts being grey.

As in the Lower Carboniferous, the sandstones and siltstones

Typical goniatites of the Namurian.

34

The geography of Britain during Namurian times.

represent periods when the river delta to the north advanced southwards into the area. The marine mudstones and rarer limestones represent briefer periods of marine transgression when either the delta retreated northwards or subsidence of the sea floor locally encouraged the formation of marine deposits. The salinity of the water in the area must have fluctuated considerably throughout Millstone Grit times, the overall trend being for the salinity to decrease as the deltaic conditions began to predominate. However, it is no more than an assumption to say that the salinity decreased with the regression of the sea.

It is generally considered that the sequence of deposition from marine transgression to deltaic transgression would commence with the formation of a marine band composed of mudstone or limestone. This would pass upwards into non-marine mudstones or siltstones as the area came under the increasing influence of the river delta which advanced into the area. This sequence would continue with the formation of sandstones and gritstones. The sandstones would represent accumulations of sand and silt heaped up in the river delta and not surprisingly in some instances, the conclusive phase of the cycle was the formation of a thin coal seam. Different parts of the river delta would encourage the formation of a variety of rock types. Indeed, although coarse sandstones were the order of the day at the height of any deltaic transgression, it is not unusual to find the coarse sandstones containing great quantities of pebbles. These were probably deposited at the bottom of the river channels.

The origin of the clastic material brought into the area by the river delta is undoubtedly away to the north on the Northern Continent. Some of the sand was made up of material rich in feldspar, which gave rise to a dark reddish colouration in the resulting sandstones. Geologically speaking, feldspar weathers rapidly when exposed to the forces of erosion and as a result, it can be said that sandstones which still exhibit the feldspar colouration today were formed at the end of a rapid process of erosion, redistribution and compaction. Grits or sandstones which do not have the feldspar colouration are indicative of slower rates of formation; the feldspar content having decomposed prior to the formation of the rock.

Geologists have divided the Upper Carboniferous (Silesian) into three divisions, the series beginning with the Namurian. The Namurian has itself been divided into seven stages which over the years have been described by the name of the most prominent rock of the times, the Millstone Grit.

The first stage in the Namurian Series is known as the Pendelian and begins with the appearance of the goniatite, Cravenoceras Leion amongst the Upper Bowland Shales. The Lower Bowland Shales are the last rock formation ascribed to the Lower

Carboniferous. The Upper Bowland Shales are visually distinct from the Lower series due to their thinner bedding. The shales extend northwards beyond the Middle Craven Fault on to the Askrigg Block, overlying the limestones, but are mainly confined to the basins. On the block, the sandstones of the Millstone Grit rest unconformably on limestone reefs of the Lower Carboniferous. Crinoid remains form a prominent part of a number of limestone beds of this age near Skipton.

Transgression by the river delta deposits are first found in the higher regions of this stage and resulted in the formation of the Newton Fells Gritstone near Settle. Over 400 metres of sandstone form the Pendle Gritstone which marks the major deposition of sandstone in the southern area.

Stratigraphical Divisions of the Upper Carboniferous (Silesian)

Series	Stage	
Stephanian		
Westphalian	D C B A	Coal Measures Times
Namurian	Yeadonian Marsdenian Kinderscoutian Alportian Chokerian Arnsbergian Pendelian	Millstone Grit Times

The second stage of the Namurian is known as the Arnsbergian Stage and is heralded by the presence of the goniatite Cravenoceras Cowlingense. Mudstones were the main deposits laid down during this stage in the basin. Sandstones and coal seams occur and a relationship is said to exist between the Tan Hill coal of Swaledale, the Edge coal of Skipton and another seam which lies between gritstone in the Pendle area. Correlation between the various beds of the different area is never easy in rocks of the Namurian, chiefly due to the complex and ever changing nature of the delta system which invaded the area during these times.

The Red Scar Grit of the Pateley Bridge area is very rich in crinoid remains, but this phenomenon is confined to the upper reaches of the grit. The Lower Follifoot Grit forms the top bed of the Arnsbergian succession. The Arnsbergian stage is followed by the Chokerian and Alportian stages but neither of these is prominent in Yorkshire, although under the name Sabdenian they achieve over 100 metres of sandstones and shales in other parts of the Pennines. The trend is for them to become thinner towards the south.

In Yorkshire these stages achieve a maximum thickness of little more than 35 metres of mudstones in and around Wharfedale. Beds of ganister, coal and limestone represent these stages in Colsterdale and indicate that water and land conditions existed at various times. This is particularly true when one considers the 25 metres of siltstones and mudstones which occur in the Cowling district. The Upper Follifoot Grit, which is really a sandstone, tops the succession and is said to be nowhere more than 20 metres thick.

The next stage in the Namurian is known as the Kinderscoutian which takes its name from the famous North Derbyshire hill called Kinderscout. The hill is capped by rocks of the Millstone Grit. Beds formed during this stage outcrop over quite a wide area, but are at their thickest in The Peak District achieving around 480 metres. In the Yorkshire Dales, especially the northern dales, the thickness of these beds is little more than 30 metres. Mudstones of the Sabden Shales series continue to be the main deposit in the basin area, but sandstones occur quite widely around Nidderdale and in Wharfedale.

The upper part of the Kinderscoutian represents another advance by the river delta from the north. The evidence of this advance is mainly to be found in South Yorkshire and The Peak District, although coarse siltstones known as Todmorden Shales are said to represent deposits from the forward slope of this advance. The lower Kinderscout Grit does not extend to the north of Leeds although it exists up to 300 metres thick in the Peak District. However the Upper Kinderscout Grit does continue to the north from Derbyshire and becomes first the Brimham Grit and then subsequently the Coomb Hill Grit near Pateley Bridge. Although the Upper Kinderscoutian is known as a period of deltaic advance, it also bears witness to no fewer than five or possibly six minor transgressions by the sea. The resulting marine bands are frail to say the least, but are irrefutable evidence of the continuing regression-transition cycle in the basin area.

We can now picture a Yorkshire landscape dominated by a river delta. The river channels are carrying large quantities of silt into the basinal area which was becoming choked with sediments. The sea was continuing to reassert itself periodically and sea floor subsidence locally also encouraged the formation of marine deposits.

Of all the stages of the Namurian, the Marsdenian most ideally indicates the cyclic character of the sedimentary deposition of the series. The record of the rocks indicates that on no fewer than four occasions the apparently ideal cycle took place. In other words, on four occasions the sea advanced back into the delta encouraging the formation of marine mudstones. As the delta retreated, the vegetation was overcome by sandstones, siltstones and subsequently mudstones, formed as a result of the sea water advance. The river delta then began to move forward resulting in the deposition of siltstones and later the coarse sandstones.

Unfortunately this ideal sequence of events is restricted in the area it covers. A depth of 200 metres is achieved in West Yorkshire, but this thins out considerably to both north and south. At Kirkby Malzeard in North Yorkshire, the Marsdenian Beds achieve little more than twelve metres and their existence in the south is mainly confined to the Ashover Grit in Derbyshire.

The final phase of the Namurian is known as the Yeadonian and is essentially a continuation of the Marsdenian. It is another phase of cyclic sedimentation. Two main marine episodes are preserved within the rocks, both of which were widespread throughout the basinal area. The major sandstone of this stage is known as the Rough Rock and covers a very wide area of country. Indeed this rock, together with the Huddersfield White Rock of the previous stage are the lowest rocks of the Namurian to cover the entire Yorkshire outcrop of the Millstone Grit. Coal bands occur within the succession and have been considered to be of workable quality and quantity in several places. The working has usually been by open cast techniques.

Volcanic rocks have no place in the Carboniferous rocks of Yorkshire, but throughout this period of geological history volcanoes had been extremely active in Scotland. In Derbyshire volcanic rocks had also been formed during the Lower Carboniferous and in a bore hole in Nottinghamshire, volcanic rocks of Namurian age have been identified.

The events I have described so far and resulting deltaic deposits, relate to the sedimentation in the basinal areas which lay to the south of the Middle Craven Fault. During later stages of the Namurian sedimentation, the basin deposits began to reach beyond the fault system on to the southern edge of the Askrigg Block. However, while the coarse deltaic sandstones and marine mudstones were being deposited in the basins, the deposits being laid down on the block in the north were rather different in character and represent a strongly contrasting environment.

The Yoredale Rocks which were being deposited at the end

39

of the Lower Carboniferous on the fault block areas continued into the Namurian. Limestones continued to be laid down in very shallow water conditions and were interspersed by mudstones, siltstones and sandstones in the cyclic pattern indicating marine and deltaic transgression and regression. However as the Namurian continued, the formation of limestones was becoming a less dominant part of the cycles. At the same time sandstones grew in importance. The lower limestones around Wensleydale and Richmond frequently are associated with beds of chert. It is envisaged that the silica rock was derived from the sponges which are a common feature of these beds.

The limestones are at their thickest on the Alston Block and show a marked tendency to become thin towards the south. The sandstones, on the other hand, increase in thickness in the same direction. The Stainmore Trough which to some extent separated the Alston and Askrigg Blocks reflects, not the shale type sediment normally associated with basin deposition, but a more complete succession of the rocks found on the blocks, and as a result offers a more complete story of the changing events of the time. The reason the basinal sediments are absent from the trough is probably because it was not only a narrow trough and, although deep, was not a basin. At the end of the Lower Carboniferous, uplift was taking place of the southern end of the Askrigg Block and this continued into the Pendelian resulting in an incomplete succession in the rocks of this age in the area.

At this point the reader may well ask an obvious if somewhat daunting question, that being, "How did copious deltaic deposits from a northern continent travel southwards into a basin, leaving apparently no sign of their passing through a block area immediately to the north of the basin?" This would appear to be an unlikely state of affairs to say the least.

We have to accept that most of the coarse, thickly bedded sandstones formed in the basin have a corresponding but much more thinly bedded sandstone to represent them on the block, consequently the deltaic deposits were influencing both areas at the same time, but in different ways. It now seems possible that the river delta approached not from the north but from the north east and as a result, deltaic deposits were restricted in their northerly deposition. Possibly the Southern Uplands of Scotland were a low lying tract of land free from the water at this time. Certainly the presence of such a land mass would greatly affect the distribution of the deltaic deposits, while at the same time being low lying, would contribute little in the way of sediment itself.

The Northern Continent was undoubtedly the source area for the deltaic sediments but their passage to the Bowland Basin must have been from the north east. Unfortunately, the channels which would bear witness to their passing probably lie across north east Yorkshire where they are hidden beneath much younger sedimentary rocks. It is also worth remembering that the northern continent was much larger than Scotland is today, stretching as it did from Ireland through Northern Britain across the North Sea to Scandinavia and beyond. We should not therefore think just in terms of the deltaic sediments coming from the area of Scotland as it is today, but possibly from an area which now forms the North Sea to the east.

Outside Yorkshire, and to the south of the Bowland Basin, similar sedimentary rocks were being formed, but the sediments did not contain the high feldspar content of the northern deltaic deposits. The lack of feldspar has led geologists to believe that the southern sediments were derived from a land mass to the south of the basin in the area of St. George's Land, and not from the northern continent.

The Namurian came to an end with the formation of a marine band containing the goniatite Gastrioceras subcrenatum, but no major change in the environment or the type of deposition was to mark this boundary. Now we must consider the character of the Yorkshire Dales area at this time.

The accumulation of sediments, both in the basinal areas and on the Pennine Fault block, had created a very low lying area composed mainly of mud and sand banks, swamps and shallow water areas, with fresh water and sea water lakes and lagoons. Vegetation had become reasonably diverse on land areas and spores wafted southwards by the wind were beginning to take hold of this delta area. The river delta dominated the scene, carrying varying quantities of water depending upon the seasons. At periods of flood the river would create a new maze of channels through the delta. Old channels would become choked with debris, new ones would carry the water away. The tropical Carboniferous Sea, where limestone had once been formed, was gone, or at least it had been forced to retreat southwards. Never again would the Carboniferous Sea dominate the Pennine arena. True, marine transgressions would still occur, but no great thickness of marine sediments would be formed during Upper Carboniferous times. Even the volcanoes of Nottinghamshire were apparently quiet by now.

The coal forests which would come to dominate the scenery during the next few million years had already begun to appear on the scene. Indeed the thin coal seams of the Lower Carboniferous age and those of the Namurian had proved the ability of the coal forests to form. The time for prototypes was at an end, the near future would see high forests made up of very strange trees, towering high into the Yorkshire skies.

5.   Coal Measures Times

THE PART of geological history following the Namurian is popularly known as the Coal Measures, but to geologists the Coal Measures are divided into five parts. The older, or lower part, is called the Westphalian A, the subsequent parts being called the Westphalian B, C and D. The series terminates with the Stephanian, but rocks of this and the Westphalian D are not known to occur in Yorkshire.

In most cases the Westphalian deposits resemble those of the Namurian, being cyclic in form, but the coarse sandstones which typified the Namurian cycles are much less important. In the Westphalian, pride of place belongs undoubtedly to coal. However equally important are the seatearths which are usually full of rootlets from the plants and trees. Not surprisingly the succession through a cycle is envisaged as mudstones (these were sometimes marine and occasionally non-marine), siltstones, sandstones, seatearths and coal. Obviously the seatearths are essentially the fossil soils upon which the coal forests thrived.

Most of the fossil content of the Coal Measure cycle is contained within the mudstones, which vary in colour from black to grey. They pass upwards into the siltstones, but there is a considerable degree of overlap between the two. Indeed one could well describe the upper reaches of the mudstones as silty mudstones and the lower regions of the siltstones as muddy siltstones. Sandstones follow the siltstones but the two are frequently interbedded. Rarely are the sandstones of a coarse character, being mainly fine grained. The sandstones are usually brown in colour when weathered and can be distinguished from the siltstones, as the latter are normally grey.

Coal generally completes the cycle and coal seams are therefore common throughout the succession, but unfortunately only a few of the seams are really worth commercial extraction. Cannal coal also occurs, but its formation is considered to be due to the accumulation of organic material in small lakes. This organic debris is not therefore entombed where it grew, but is nothing more than an organic detritus.

Ironstones occur commonly throughout the Coal Measures

The geography of Britain during Westphalian times.

although they are not a feature of the coal itself. In the form of siderite they frequently occur as concretions which, on being broken, are found to contain small veins of calcite. In some cases the ironstones have been worked for commercial purposes as a source of iron ore. The seatearths are also occasionally of economic importance. The leaching out of minerals from where a clay soil existed may leave a grey fireclay which is used in the manufacture of refractories. Ganister, which is a pure siliceous sandstone, also occurs as seatearths of the type derived from a sandy soil. Ganister has much the same commercial importance as fireclay, but may be a more economically viable material.

The fossils which typify the rocks of the Coal Measures are the non-marine mussels (bivalves). They occur as mussel bands at several different horizons within the Coal Measures. Individual genera tend to become dominant at different times throughout the succession and these have been used to help characterise the different stages within the Coal Measures.

Repeated incursions by the sea resulted in the formation of marine bands which contain other fossils, including goniatites which were used to classify the Visean and Namurian stages. Goniatites have again been used to this end in the Coal Measures. Fossil plants have also been used to characterise this succession, the sequence being divided into nine zones which have been

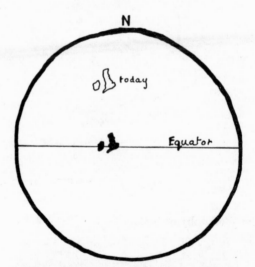

The position of the British Isles during
Westphalian times was close to the equator.

given letters from A to I.

Unfortunately the subsequent folding and erosion of the coal strata in the Pennine area means that in the Yorkshire Dales they are very poorly represented. Indeed the rocks of the Coal Measures are confined to the area around Ingleton in the west and two small areas north of Pateley Bridge. Consequently our knowledge of the geological history of the Yorkshire Dales is incomplete. However, our knowledge of the exceedingly small Ingleton coalfield combined with the understanding we have of the infinitely larger Yorkshire, Derbyshire and Nottinghamshire coalfields, has enabled an approximate picture of happenings in the Yorkshire Dales to be made. It may come as a surprise to some readers that the geological events in other countries can be used to piece together events which took place in a small part of Yorkshire. However it is important to remember that the coal forming conditions existed over a large area of Europe and Asia at the same time. In fact coal deposits were being formed from Ireland, through England, across the North Sea, along the Baltic, through Germany and Russia, at least as far as the Ukraine. In the face of such an enormous area the Yorkshire Dales appear to be insignificant indeed.

WESTPHALIAN A

At the end of the Namurian the Yorkshire scenery was one of a low lying area made up of mudbanks and sand banks of a river delta. The climate was warm and the area was becoming subject to swamp conditions. The land, such as it was, never rose far above the waters and the waters were never to submerge the land by any significant depth. Vegetation was beginning to increase both in quantity and in variety, and plants were starting to take hold on the mud banks. Spores were wafted southwards from the Northern Continent and slowly but surely these germinated wherever and whenever they were given the chance.

The Westphalian A began with the Pot Clay Marine band, but this was only the indication that there would be no fewer than six major marine transgressions during this stage. In fact these transgressions by the sea completely dominated the lower part of the succession. Although cycles were taking place, many were to be incomplete. The formation of coal bands did take place, but without exception they are thin. Sandstones on the other hand were reasonably prominent.

Coal bands are few and far between amongst the rocks of the Westphalian A, but the upper part of the sequence taken in the Barnsley area of the Yorkshire coalfield contains no fewer than twelve coals. It is interesting that although sea water conditions invaded the area many times early in the succession, this trend

45

decreased with the passage of time. Coal formation on the other hand became more common as the marine conditions decreased. Mussel bands are quite common throughout the succession.

We can now picture the area becoming increasingly under the influence of the coal forests. In some areas sea floor subsidence or a rise in sea level allowed the sea to infiltrate back into the delta area. However, the tendency for marine conditions to return became less common as time progressed. We can also imagine heavy rain encouraging the erosion of the northern continent and the rivers becoming periodically swollen with flood waters. Large quantities of river debris would be carried down to the delta by the river, which would carve new channels out of the swamp as old ones became silted up. The time of the great coal forests had come. Trees, which to us in the 20th century would appear very unusual, were by this time beginning to tower up to 100 feet into the sky.

Considering that the Carboniferous coal forests were formed around 300 million years ago it is hardly surprising that the trees were very different from those we know today. What may seem surprising is that the trees were relatives of several plants which still inhabit the planet. The Calamites were particularly abundant, these were primitive horsetails. Club mosses were represented by the Lepidodendron and the Sigillaria, but there was also a wide variety of ferns. All these plants were spore rather than seed bearing. The fossil horsetails frequently form the basis of the coal deposits and had roots up to 40 feet in length. The roots and rootlets are known as Stigmaria.

WESTPHALIAN B

The lower limit of the Westphalian B is represented by the base of the Clay Cross Marine Band. The lower part of this stage is very similar to the upper reaches of the previous stage and the Clay Cross Marine Band represents the only early transgression by the sea. The formation of coal seams, many of which are or have been worked commercially, was at its height. Sandstones were not prominent, the main ones occurring in the vicinity of the Clay Cross Marine Band. Mussel beds continued to occur frequently. Undoubtedly the Yorkshire scenery was one of a low lying swamp.

The upper part of this stage saw a return of the sea which resulted in the formation of many marine bands. Predictably this time of marine reassertion was at the expense of coal and in fact few of the coals near the top of the Westphalian B have been worked commercially. Sandstones also come back to the fore and one, the Woolley Edge Rock, has been worked extensively for building purposes. The top of the Westphalian B stage is represented by the base of the Mansfield Marine Band which was the most important advance by marine conditions during Westphalian times. The rocks of the Ingleton coalfield are believed to correspond to the Westphalian A and B stages.

WESTPHALIAN C

The Mansfield Marine Band represents a major invasion of the swamp area by sea water and indicates that the marine conditions persisted for some considerable time. Indeed the marine band is believed to exceed ten metres in thickness in some areas. Following the marine band, cyclic conditions again prevailed and a large number of coal seams resulted. Sandstones are an important feature of the succession but are frequently interbedded with siltstones. None of the sandstones are coarse in character. Many of the coals are thick enough to have been worked commercially, but only two have so far been exploited in the Yorkshire coalfield.

Several marine bands occur in the lower part of the Westphalian C succession and the final band of this stage in Yorkshire is called, not surprisingly, the Top Marine Band. From this point onwards the coals thin out and are few and unimportant. Sandstones become dominant and are occasionally quite coarse. Several sandstones have been worked for commercial purposes.

Mussel bands are numerous towards the top of the Westphalian C and have been used as marker horizons in the upper reaches of the succession. The fossil content of the higher mussel bands has left open the possibility that the very top of the sequence in Yorkshire may possibly belong to the Westphalian D. However, this is no more than a possibility. The uppermost beds in the succession are red in colour and consequently are called Red Beds. This colouration is thought to be due to the dry oxidising conditions of the Permian desert landscape which were, 270 million years ago, waiting in the wings of the earth's stage.

WESTPHALIAN ROCKS IN THE YORKSHIRE DALES

The Ingleton coalfield is small in comparison with the major coalfield in Yorkshire and measures little more than 60 square kilometres. No mining has been carried out since the 1930s. The rocks in this area correspond to the Westphalian A and lower parts of the B succession. Six coal seams have been worked commercially, two of which are attributed to the lower part of the Westphalian B stage. Much of the Westphalian B is missing due to erosion and the coal bearing rocks are overlain by the Red Beds which correspond to the Westphalian C. The Red Beds at

Ingleton are composed of conglomerates, breccias, sandstones, several limestones, ironstones, and shale. Coals are totally absent from the succession.

Rocks of the Coal Measures are known near Ripon. Shale and sandstone make up most of the succession which is considered to belong to the base of the Westphalian A stage.

In Yorkshire the Red Beds mark the end of the Carboniferous Period, but this is only on account of the fact that the youngest or uppermost rocks, those formed during the Westphalian D and Stephanian stages, were removed after their formation by the forces of erosion.

We can now picture a landscape in the Yorkshire area where gentle pressure is beginning to throw the Carboniferous strata into undulating folds. Britain has by now moved to a position just north of the equator and is a central part of a huge continent. The dry climate would soon combine with erosion to create desert conditions similar to those in the Sahara today. The river delta and coal swamps would soon pass away leaving only the rocks to bear witness to their passing.

Approach to Malham Cove, a dry waterfall in the Great Scar Limestone.
[Photo: J. R. Bunting]

Water Sinks. The little girl's feet are immediately above the spot where the Malham Water disappears into the limestone, to reappear not at Malham Cove but at Aire Head.

Gordale Scar from the air. This famous location may have been a cavern, the roof having collapsed, or it may have been created by waterfall recession. [Photo: C. H. Wood]

The Great Scar Limestone, beautifully displayed at White Scar near Ingleton. The scene is overshadowed by the towering mass of Ingleborough. [Photo: Tom Parker]

Thornton Force. The junction between the old Ingletonian rocks and the Great Scar Limestone can be clearly seen part way up the falls.

The entrance to White Scar Cave. The Great Scar Limestone forms the roof of the entrance while the Ingletonian rocks form the floor and walls.

Sedgwick House, the birthplace in Dent of Adam Sedgwick the famous geologist. The inscribed monument is in Shap granite.

North of Sedbergh pre-Carboniferous rocks make up the scenery. The hill, bottom left, is called Bluecastor and is partly formed out of an igneous intrusion.

Wensleydale, once known as Yoredale. The stepped hillside in the photograph is evidence of the Yoredale rocks of which it is composed.

The Old Gang Mine near Swaledale is just one of many relics of a forgotten mining industry.

The remains of the old colliery at Tan Hill. The coal was taken for use in the smelting mills in Swaledale more than a century ago.

Galena crystal on geothite, Arkengarthdale.

The church at Horton in Ribblesdale is overshadowed by Pen-y-ghent, which is capped with Millstone Grit.

6. Bringing Events up to Date

THE Carboniferous Period of geological history was brought to a close by the onset of the main crustal movements of the Hercynian Orogeny, when mountains were being created away to the south east. Britain was very much on the fringe of this happening, but it did have an effect upon the landscape. Although the effect of the Hercynian Earth Movements was only slight in Britain, it did result in the rocks of the Pennine area being folded and uplifted well above sea level. Immediately erosion set in and the destruction of the rock surfaces began.

The Upper Carboniferous rocks, which were in effect the Coal Measures, were the first to be attacked as they lay upon the top of the pile. It is anticipated that the softer rocks of the Coal Measures were quite quickly removed, exposing the coarser and thicker sandstones (gritstones) of the Namurian. Erosion of the Carboniferous rocks was to continue for around 50 million years before it would come to an end. By this time Britain was situated close to the centre of a super continent which geologists have called Pangea. Since it was many miles from the sea and close to the equator, the country became the scene of hot arid and desert like conditions.

As the erosion of the Carboniferous rocks continued the limestones of the Dinantian or Lower Carboniferous era were exposed below their sandstone cover. Of course the Carboniferous rocks had been thrown into a number of gentle folds and it was the top of the folds which were subject to maximum erosion. This is why in some areas the softer rocks of the Coal Measures have been preserved. The Coal Measures of today are those which were low lying and were protected from the full effects of erosion.

The forces of erosion have but one end in view, this being to reduce the land above sea level down to sea level. A land surface which has been subject to erosion for a very long period of time may become extremely low lying and featureless. Such an eroded land surface is known as a peneplain. Although at one stage the Carboniferous rocks of the Pennines may well have been thrust upwards into hilly or even mountainous terrain, a peneplain surface was more or less achieved by the end of the period of erosion. However, it is envisaged that the harder and resistant limestones of the Lower Carboniferous, and the sandstones of the Namurian period would still have maintained some degree of relief on the landscape.

It is now possible to imagine Britain lying in the shadow of a huge mountain chain which existed to the south east. Desert conditions prevailed and as the mountains were eroded, the plains were accumulating the debris. Slowly but surely the eroded peneplain of Carboniferous rocks was covered by sand and rock debris. To the east, the North Sea Basin was being created and the Zechstein Sea would soon grow and spread from the east to cover parts of Central Yorkshire. Once buried beneath the desert sediment, the Carboniferous rocks were protected from further erosion. However, the shaping they had already received would form the basis of the Pennine landscape we enjoy today.

The late Carboniferous early Permian part of Yorkshire's geological history is also notable for the formation of many mineral veins and the Whin Sill. The sill, which now falls outside the county boundary due to the modification of local government boundaries in 1974, took many millions of years to form. It is at its thickest in Teesdale, but traverses a large area of the North East before apparently terminating at the Farne Islands off the Northumberland coast.

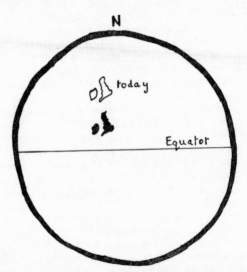

The position of the British Isles on the earth's surface by Permian times was well north of the equator.

The geography of Britain during the latter part of the
Permian period.

The mineralisation must have taken place after the formation of the sill, for in fact the sill is affected by the mineralisation. That the sill took many millions of years to form is also a clue to the structure of the mineral veins, for in some cases they also took long periods of time to develop. However, while the mineral veins of the Alston Block were emplaced during the late Carboniferous, early Permian times, in other parts of the Pennines, notably the Peak District, many of the mineral veins were formed much later. Certain veins have actually been dated around 180 million years old which means their formation took place during the Jurassic Period of geological history.

With the burial of the Carboniferous rocks, many events would take place which would leave little or no trace of their passing upon the landscape we know today. The Lower Permian is represented by breccias and sandstones in other parts of Yorkshire, but with the flooding of the North Sea Basin during Upper Permian times, marl and the famous Magnesian Limestone were formed.

Later, repeated evaporation of the waters of the Zechstein Sea would create large quantities of evaporites which include such commercially desirable minerals as halite and anhydrite. The final rock formed during Permian times was sandstone, but there is still some disagreement as to where the Permian ends and the next period, the Triassic, begins; for sedimentation was continuous in the North Sea Basin from one period to the other.

Life had continued to evolve and the Permian Period saw the arrival of the first reptiles, although these were mainly small. The amphibians had continued to evolve and were much more numerous. Aquatic reptiles, including the needle-toothed Mesosaurus, were also thriving in the seas and were paving the way for the huge reptiles which would later dominate the earth for almost 200 million years; the dinosaurs.

The Triassic Period, which began around 225 million years ago, followed much the same lines as the Permian. In Yorkshire the Triassic rocks were essentially formed around the western perimeter of the North Sea Basin. Pebble beds formed much of the sediment during early Triassic times. The pebbles which were mainly quartzite are believed to have had a southern source. Sandstones followed the pebble beds, which were in turn followed by the Keuper Marl. The marl, which was essentially a calcareous mudstone, also contains beds of sandstone, and evaporites such as anhydrite and gypsum. These were formed by the repeated evaporation of water, rich in the minerals concerned.

Subsidence of the land surface towards the end of the Triassic allowed the sea to encroach upon the land, and as subsidence continued, the sea finally flooded the entire area. However, subsidence and uplift of the sea floor was a frequent happening during the next 60 million years of the Jurassic Period as it had been 150 million years earlier when the Carboniferous Sea had covered the area.

By now the super continent Pangea was breaking up and this was causing the subsidence of the earth's crust. The giant reptiles which we call dinosaurs were dominating the surface of the earth on land, sea and in the air. Large herbivorous dinosaurs which weighed many tons roamed the fringes of the seas and lakes taking support from the waters for their vast bulk, and obtaining sustenance from the thick vegetation. Small but less likeable carnivorous dinosaurs prowled the landscape making their meals off the plumper herbivores. Aquatic reptiles had become the scourge of the oceans, and in the air the flap of mighty bat-like wings would herald the approach of an aerial reptile intent on finding its prey.

The Jurassic Sea covered much of England including the Pennines, the shore being somewhere away to the west. Shales and sandstones begin the Jurassic succession but the waters, which at first were muddy, became clear and Oolitic Limestones were formed. Later movements of the sea floor caused the land in places to rise above sea level.

Vegetation abounded upon these long areas of land and the reptiles were increasingly numerous and diverse. Trees were the principal form of vegetation for grasses had not yet appeared upon the scene. Insects were many and varied in character, and included dragonflies which had now existed for some time. Birds had not yet put in an appearance, but the bird-like reptiles would soon begin to resemble reptile-like birds.

The Jurassic Period came to an end as the earth's crust in the area of Central and Northern England began to subside more or less uniformly. The seas encroached upon, and then covered the land, only the Highland area of Scotland and parts of Wales remained above water.

This event heralded the dawn of the Cretaceous Period of geological history. The sea water was warm and teamed with life. Indeed this was the heyday of such creatures as the ammonite

The ammonites were descendants of the goniatites which thrived during the Carboniferous.

which evolved to tremendous proportions. The ammonite had descended from an earlier genera called goniatites which have been used by geologists to classify the rocks of the Carboniferous system in Yorkshire and elsewhere. However, now the ammonites were nearing extinction.

There was no important land mass in the area and the waters of the sea were crystal clear. Slowly over a period of time, the lime-rich shells and skeletons of the myriads of small creatures living in the waters were built up on the sea floor. They formed a deep deposit of white mud. This white mud was one day to form the chalk which dominates the county of Humberside today. Chalk, which is a pure limestone, also underlies parts of Lincolnshire, East Anglia and South East England. The white deposit deepened and solidified into chalk, but the earth was on the move yet again.

Slowly the sea bed began to rise 'from out the azure main'. This event, which marked the end of the Cretaceous or Chalk Period, also marked the beginning of a new phase in Yorkshire's geological history. The Carboniferous rocks of Yorkshire had now been hidden for over 100 million years beneath an ever-increasing cover of sandstones, limestones, shales and chalk. Now the process was to go into reverse. However, the end of the Cretaceous Period marked not only the start of a new phase in our geology, but also the end of a way of life.

Life on the planet earth has been continually evolving for around 2,000 million years. By the end of the Cretaceous era the giant reptiles we call dinosaurs had evolved to a sophisticated level and they were masters of air, land and sea. Undoubtedly they were the most advanced life form at the time, but they had become specialised to a high degree. For over 200 million years they had dominated the world but the end was imminent.

Fossils of the dinosaurs are abundant in rocks of Cretaceous Age, but are nowhere known in younger formations of Tertiary Age. For in fact the end of the Cretaceous Age was also the end of time for the dinosaurs. But what tumultuous event brought about their almost total demise? Many theories have been put forward to account for the passing of the reptiles, but this is one subject around which controversy still rages. One theory is that the small mammals, which were by now fairly numerous, ate the dinosaurs' eggs. Be that as it may, a more likely answer lies in understanding the weather.

It is generally envisaged that towards the end of the Cretaceous Period the climate began to cool. This, over a long period, led to a change in the nature of the vegetation. The copious foliage of the plant life the herbivorous dinosaurs ate, began to disappear. When the large dinosaurs began to diminish in numbers, the small carnivores which lived off them also had difficulty finding food.

The survival of the dinosaurs may well have hung on a thread, but it seems unlikely that this environmental change should have in itself made the entire species extinct. It is interesting to note that many other creatures including the ammonites became extinct at this time.

There is also evidence to suggest that a more rapid change brought about their extinction. Such a change would be accounted for by a sudden rise in the sea level. Whatever the cause of their passing, the dawn of the Tertiary saw the world free from the giant reptiles and at last the small mammals were able to come to the fore, and prepare the way for the ascent of man. It is interesting to wonder how man would fare if confronted with the same catastrophe which overtook the dinosaurs, whatever it may have been.

By the dawn of the Tertiary Period, Britain was situated well to the north of the equator. Most of Britain had been under water during the Cretaceous Period, and the Tertiary was heralded in by an uplift of the sea floor. The sea floor was uplifted out of the sea and became dry land. The entire area was not uplifted uniformly, but was generally centred around the Pennine area. The landscape had the appearance of an elongated dome. Uplift continued and eventually a large island was formed with a chalk landscape.

Away to the north, volcanoes were active between Greenland and the islands of the Hebrides, due to the opening up of the Atlantic Ocean as Europe and America parted. To the south east the Alpine Orogeny was getting underway and this would result in the uplift of south east England.

The chalk landscape of the new Britain was immediately subject to erosion, and rivers began to make channels in the white landscape. Before long the chalk was removed from the hilly areas revealing the Jurassic rocks which lay below. The landscape was changing fast and true flowering plants had finally arrived upon the scene. Grasses were also present and birds flew in the sky. Of the dinosaurs there was no sign.

Erosion pressed ahead and the Jurassic rocks were slowly removed, exposing the younger Triassic and Permian rocks to the eye. Finally the old sandstones, limestones and shales of the Carboniferous era were exposed to the atmosphere for the first time in 120 million years. At first the younger rocks still remained on the lower slopes of the central hills, but time was to see these totally removed. The Carboniferous rocks were the hardest erosion had met to date and the rate of erosion decreased. By now the landscape was beginning to resemble the structure we know today.

No one knows how long it took the forces of erosion to wear away the younger rocks, but another highly erosive force was about to show its hand. The time is two million years ago. Slowly but surely as the years passed the climate became cooler. Year by year the winters became colder. The winter snows which accumulated on the hills and mountains of Britain were not melted by the heat of the summer sun, and so white clad peaks became a permanent feature of the British scene. As the years passed, each being slightly colder than the one before, the snow accumulated on the high ground until eventually it became compacted into glacier ice and gravity began to draw it off the mountains. The icy tentacles of the glaciers began to feel their way out and over the British countryside. The glaciers united as they reached the lower ground and became Ice Sheets. The most mountainous areas of Britain are in the north and consequently the Ice Sheets moved southwards over a British tundra landscape.

The ice ground its way over the landscape. In places it was several hundred feet thick and it probably covered many of the hills of southern Britain. As the ice moved, it abraded the landscape breaking and shaping the exposed rocks to its will. Instead of smoothing and rounding the hills over, the effect of the passage of the ice was to sharpen the scenery, making the hills resemble mountains.

Hundreds of feet of ice moving over the landscape must have a tremendous effect upon the scenery and indeed the Ice Age was to provide the finishing touches to the British scene. However, the effects of the ice were not confined to modifying the existing scenery, it also added a new 'rock'. As the ice moved along over the countryside, it accumulated great quantities of rock debris beneath it. The debris was made up of soil, mud and rock fragments which had been broken away from existing rock surfaces. The debris was carried southwards under the ice when ultimately

The maximum advance of the ice sheets over Britain never reached south of a line from the River Severn to the Thames.

the ice melted, the debris which is now called boulder clay was left behind, covering the rocks which had previously been exposed. Boulder clay is easily identified because of the fact that it is a clay matrix with pebbles and rock fragments protruding from it. It is soft and consequently it was rapidly eroded from the higher ground. However, the ice had not yet melted and many thousands of years would pass away before the clay would be exposed.

The first advance of the ice over the British scene began approximately one million years ago. The ice sheets advanced far to the south of Yorkshire to a latitude where the ice could not survive the warmer climate. In effect this was a snow line, beyond which snow would fall, but would be melted by the heat of the summer sun.

Slowly the climate became warmer and the ice began to recede northwards leaving the boulder clay behind. This was the start of an interglacial period when a mild temperate climate would return to Britain. The forces of erosion went to work on the soft boulder clay and quickly the older rocks were exposed again.

During the last million years, the ice has advanced southwards over Britain on no fewer than three occasions. Each time the ice has created havoc on the rock scenery and each glacial period has been terminated by a warm interglacial period. The ice worked its way southwards, but never crossed to the south of a line between the River Thames and the River Severn. The third glacial advance came to an end around 10,000 years ago. By this time man was beginning to make some impression on the scene.

The boulder clay on the high ground of the Pennines was quickly eroded. Indeed the clay would only be preserved on the lower ground which flanks the Pennine Hills. And so the Carboniferous rocks were once again exposed to the elements and their erosion continues. The final shaping of the Pennines, and for that matter the rest of Britain, had taken place and the scenery was very much the way we know it today.

Is this the end of the story? Unfortunately not, for we are only living through another interglacial period and at some time in the future the ice will once again move southwards over Britain. Volcanoes and earthquakes which, at the moment are not a feature of our country, will return and Britain will hear the once familiar roar of volcanoes yet again. Erosion will continue to grind the hills and mountains down to sea level, while the uplift of southern England will continue.

This series of events may provide a stimulating, if not tempestuous future for our country, but like all geological happenings it will be 50 or 100 million years before they come to pass. Such happenings will not provide us with any discomfort.

7.

Lead Mining in the Yorkshire Dales

TODAY, apart from limestone quarrying, the beautiful dales of Yorkshire are free from any major industry other than agriculture. This I can say with an intense feeling of relief because wherever industry encroaches, the beauty of the scenery is very much at risk. Of course the Yorkshire Dales are the heart of a national park, and hopefully industry in a 20th century sense will never gain a foothold. However, such has not always been the case.

Despite the fact that the Yorkshire Dales should not experience any large scale industry again, the very beauty of the Dales is, in some almost indefinable way, enhanced by the waste materials left over from a bygone mining industry — an industry which reached its peak during late 18th and early 19th centuries. It was the Industrial Revolution which created the need for the extraction of metal ores to take place on an ever-increasing scale. Indeed so great was the demand for Yorkshire's mineral wealth that mine workings grew deeper and deeper and the waste heaps on the surface grew larger and larger. Before long the workings had become so extensive that all the signs left behind by over 1,500 years of mining were obliterated from the scene.

Mining for galena, the principal ore of lead, first began in Yorkshire before the days of Roman Britain, but unfortunately we shall never know just how early this activity began. There is however no doubt at all that the earliest mining did no more than scratch the surface of the mineral veins. The most primitive techniques which would have been available to the miners of two thousand years ago would have eliminated all but the small scale open cast approach. The Brigantes, an Iron Age people who claimed the Yorkshire area as part of their territory, were definitely extracting lead on a small scale at the beginning of the first century A.D. The Romans came and following the Battle of Stanwick Park in A.D. 74 at which the Brigantes were heavily defeated, the extraction of lead in the Yorkshire Dales passed to the victors. However there can be little doubt that the Brigantes, many of whom were now slaves, did the work under Roman supervision.

Mining continued throughout the days of Roman Britain, but at no time did the mining reach the sophisticated level that it did in other parts of the country. The concept of slaves crawling about in dimly lit tunnels extracting the lead ore as they did in Shropshire does not apply to the Yorkshire mining field. However, pigs of lead have been found in the Yorkshire Dales which could only have been of Roman origin. Each of the lead pigs carried the name of the Roman Emperor of the time and a date, presumably indicating when the pigs were produced.

The Romans used considerable quantities of lead in their buildings, but due to the fact that lead ore was available in many parts of England, the mining techniques used were probably never more than superficial. Certainly many of the other minerals which occurred with the lead sulphide had no use at this time. Indeed the 20th century would dawn before many of these minerals would become as important to man as galena was to the Romans.

When the fall of the Roman Empire was imminent, the legions were recalled from Britain and an uneasy peace reigned for a good many years. During unsettled times the extraction of the lead ore diminished and mining was only carried out spasmodically through the next few centuries. The Middle Ages saw the extraction of the ore increase dramatically, but these were also unsettled times and many minor wars would be fought over the ownership of the mineral veins. During the 12th century the right of lead extraction in various parts of the Dales passed to the two great abbeys of Jervaulx and Fountains and so it continued until the Dissolution of the Monasteries several hundred years later.

So far the extraction of lead ore over a period of a thousand and more years had done little to deplete the mineral veins, but the times were soon going to change. It was the Industrial Revolution which made the difference. All at once the extraction of lead ore became of paramount importance in a new world which was thirsty for the earth's metal ores. Mining techniques were still primitive, but at last man was going to do more than just lift the lid off the veins; now he was going to search around inside.

The dawn of the 17th century saw lead mines working in many of the Yorkshire Dales including Teesdale, Wensleydale, Nidderdale, Wharfedale, Swaledale and Arkengarthdale. And by now the miners were going deeper, as a new wealth of lead ore reached out before them. However, mining equipment had hardly progressed beyond the pick and the shovel.

Lead mining in the Yorkshire Dales as in many other areas of Britain reached its peak during the 18th century and naturally enough large quantities of ore were being removed from the earth. In fact only about ten per cent of the mineral vein was composed of the profitable ore and consequently large amounts of unwanted mineral or 'gangue' was brought out with the ore. The gangue minerals were primarily calcite, barite, and fluorite although other secondary minerals were also present. Small quantities of copper and zinc ores were extracted, but they only formed a very minor part of the vein.

On arriving at the surface, the ores were subjected to a variety of separation techniques resulting in only the desirable minerals going on to the smelt mill. In practice, large quantities of gangue minerals went to the smelter and similar quantities of ore went to the mine dumps which is where the unwanted minerals were allowed to accumulate. The 'old man', as the old lead miner is called today, had little thought for the scenery. No effort was made to put the waste minerals out of sight, they were simply tipped as close to the mine as possible. And so the mine dumps grew and grew until they stood out as great, in some cases gigantic, epitaphs to the lead mining industry. Even today when the lead mining industry has gone, and most of the old mine buildings and chimneys have disappeared from the scene, the mine dumps still dominate the Yorkshire Dales.

To return to the 18th century, many of the long established mines began to run into problems. The search for an ever-increasing quantity of lead ore meant that the mines had to go deeper into the earth. It was not long before the shafts reached the water table and production ground to a standstill. Various different types of pumping engines were utilised to pump the water out of the mines, but this was to prove to be an unending problem which was overcome with varying degrees of success. To the small mining company these problems could not be overcome financially and many were taken over by the large companies. One of the biggest companies in the north of England was the London Lead Company which, working from Middleton in Teesdale, operated successfully for almost 300 years.

The latter part of the 19th century saw the lead mining industry going into decline. Many of the mineral veins were effectively exhausted, or could not be worked deeper because of the water problem, and so the mines began to close. The falling price of lead, on account of new sources of supply entering the market, and the ever-increasing cost of ore extraction were sounding the death knell of this once profitable industry. Few mines were still operating at the beginning of this century.

And so the sound of pumping engines, sieves and crushers died away. The mine buildings, engine houses and chimneys began to fall into decay. The smelt mills were also redundant, even some of the smallest open cast coal mines in the Dales, which were supplying the mills, had to close. Work was hard to get and the days of the depression loomed ahead. The Yorkshire Dales had lost a way of life, only sheep and their shepherds would climb the Dales on their lonely vigil. However, the Dales had not seen the end of man's mining activities, it would come again, but next time it would not be lead ore the miners would be seeking.

Today mining is a small industry in the Yorkshire Dales and never again should it be allowed to dominate the scenery, even if highly desirable minerals were present. Over the years man's needs have changed, lead is no longer so desirable, having been recognised as the pollutant it undoubtedly is. Instead fluorite is being worked, as is barite; both minerals which until comparatively recently were considered to be worthless. Fluorite is used in the manufacture of steel and barite has considerable application in the paper industry and is also used as a lubricant in the North Sea oil fields. In many cases these minerals are not mined from underground, but are taken from the old mine dumps on the surface.

Mineral collecting is a very popular hobby at the moment and the old mine dumps in the Dales can offer many fine examples of the more common minerals. Permission to examine the dumps can be obtained from land owners. Under no circumstances should any collector enter the old mine workings. These are often highly dangerous and should only be entered by suitably qualified people. Even a horizontal adit (tunnel) may have vertical shafts leading off into the depths of the mine. Old rotting timbers may be all that covers a 400 foot shaft.

The minerals the collector is most likely to find on the spoil heaps are galena, barite, fluorite, calcite, sphalerite, witherite and possibly the occasional specimen of malachite. Unfortunately quartz in its attractive crystal forms is not found in the mineral veins in the Yorkshire Dales although it may be found on mine dumps in Teesdale. However, quartz in the form of chert is abundant in the Dales and has been worked on the hills to the north of Reeth in Swaledale where it can be seen littering the landscape today.

8. Geological Locations in the Dales

THIS chapter is devoted to an area by area description of the geology of the Yorkshire Dales as the visitor may visually appreciate it today. However, as I have mentioned in a previous chapter, the geology of the Yorkshire Dales does not end at the county boundary, but overspills into the counties beyond. In view of this fact and as an aid to completion, I have included two areas which are beyond the county boundary. One of these is Teesdale which for centuries was partly in Yorkshire, and the other is Bowland which is of course in Lancashire.

A problem for any author writing on the subject of geology is that of conservation. The very fact of thousands, if not millions, of visitors tramping the Dales every year creates problems of footpath and soil erosion. It may seem nothing for one person to enter a quarry and take away a specimen of the local rock or mineral, but if a thousand people visit the location and take away a thousand specimens, the location may well cease to exist. Bearing this fact in mind, the reader will no doubt excuse the fact that the following chapter tends to be general rather than precise.

Visitors to the area who intend visiting the locations described in this chapter are asked to abide by a few simple rules.

1. Always follow the Country Code.
2. Always obtain permission to enter private property and abide by any instructions given by the owner.
3. Leave the locality free from hazards such as plastic bags, tins and the like.
4. In quarries avoid walking near the working face, or on recently blasted rock which may be unstable.
5. When exploring old mine dumps beware of unguarded shafts. Many of these shafts, which are hundreds of feet deep, are still inadequately covered. One should also resist the temptation to enter mine adits (horizontal tunnels leading into mines). This is a task for professionals only.
6. Do not hammer on exposed rock surfaces.

7. Minerals can be collected from the old mine tips in the Yorkshire Dales, but please do not take more specimens than you really need.

AIREDALE

Malham
The pretty village of Malham is the gateway to some of the most fascinating limestone scenery in Britain. Several walks can be taken depending upon the agility of the person concerned. Each route highlights the various characteristics which make limestone country so special.

Janet's Foss
The most comprehensive route which takes in many interesting features begins from Malham village, but should only be taken in fine dry weather. The route lies to the east from the village. At the first road junction, turn right and proceed for several hundred yards. Opposite the derelict barn a stile gives access to the path to Janet's Foss. This is a limestone waterfall in a beautifully secluded tree-lined glade.
(N.B. Changes in the route of the above footpath are probable; details can be obtained from the Information Centre).

Gordale Scar
Returning to Gordale Lane, continue to the right over the bridge taking the gate on the left which leads to Gordale Scar. The valley is traced by a path leading towards the Scar and may be the remnant of a cavern, the roof having collapsed long ago. The waterfall with its tufa screen makes an inspiring sight, especially if the weather is dry and you can climb the falls and look down from above.

Malham Tarn
Having climbed the falls, the path continues to Malham Tarn. The waters of the Tarn wash over a bed of Silurian rocks and glacial debris. Across the Tarn stands Malham Tarn House which is now used as a Field Study Centre. Approximately $\frac{1}{2}$ mile to the west beyond the Tarn stands the remains of a smelting mill which was used for lead smelting back in the last century.

Leaving the Tarn, the path follows the Malham Water until the water disappears underground at Water Sinks. Contrary to what one would expect, this water does not flow out at the foot of Malham Cove, but reappears at Airehead to the south of Malham village. The path which from here forms part of the Pennine Way then progresses along the dry valley towards Malham. The

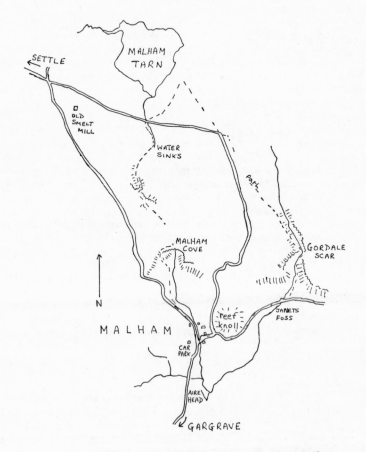

Map showing locations of geological interest in the Malham area.

Malham Water must have flowed along the limestone valley before it found another easier route underground. The limestone along the dry river bed shows typical vertical faulting created by weathering.

The same weathering is responsible for creating the clints and grikes of the limestone pavement above Malham Cove — which is a long dried up waterfall—in this case the vertical faults in the limestone have been opened up by the frost and rain water.

From here a path across the top of Malham Cove leads back to the village running alongside the Malham Beck. The Beck which flows out from the base of the Cove actually originates close to the old smelting mill near Malham Tarn where it is called Smelt Hill Beck.

South of Malham, limestones of Tournaisian Age were exposed in a number of very old quarries at Winterburn and at Bell Busk.

ARKENGARTHDALE

One of the most lovely and remote dales in Yorkshire, Arkengarthdale dissects moorland scenery which is based upon rocks of the Namurian age, generally called Millstone Grit. Exposures of rock are not common, especially in the lower reaches near Reeth, but the importance of minerals in the area cannot be denied.

From Reeth northwards the floor of the dale is founded upon alluvium and boulder clay, but on the north east side of the dale the succession consists of limestones of the Dinantian period which are capped by cherts of the Namurian. In the west, Calver Hill is made up of limestones including the Main Limestone and old quarries on the hill were worked many years ago for chert. Mine dumps become an increasing feature of the dale as we progress northwards. Indeed mineral veins dissect the hillsides on both sides of the dale and are most abundant immediately to the north of Arkle. The Great Blackside Vein cuts across Arkengarthdale approximately 500 yards beyond the hamlet of Langthwaite.

The Middle Limestone of the Dinantian period is exposed along the north side of Arkle Beck between Eskeleth and Whaw. The road to The Stang Forest cuts through the Middle and Five Yard limestones before ascending onto the Richmond Cherts. The mine dumps created during the working of Black Vein may be seen to the west of this road. The dumps contain galena, calcite and goethite.

Back in Arkengarthdale, the small track from Whaw onto Faggergill Moor leads to old mine workings which yield a variety of minerals including galena, barite and calcite. From Whaw the road climbs out of the dale and at Tan Hill joins with the minor road which climbs out of West Stones Dale.

Reeth High Moor

The minor road from Arkengarthdale (begins $\frac{1}{4}$ mile north of Langthwaite) to Swaledale offers access via several paths onto Reeth High Moor, Whaw Moor and Reeth Low Moor. The road climbs the succession from the Main Limestone through to chert and sandstones all of the Namurian type. This moorland area is the site of the Old Gang Mine (Turf Moor) which in its day was one of the most productive mines in the Pennines. Today the spoil heaps of this old mine are being reworked for fluorite which is used extensively as a flux in steel making.

Reeth High Moor is the site of a smelting mill where the lead ore, called galena, was smelted into lead. The poisonous nature of the gases evolved during the smelting process made it necessary for the smelting mills to have very long flues which traversed the countryside for considerable distances. The exhaust gases cooled as they passed through the flues, the obnoxious lead content being deposited upon the walls of the flue. This technique helped to minimise the pollution of the surrounding countryside. The flues, together with the other mill buildings, provide a fascinating insight into lead smelting during the 18th and 19th centuries. However most of the buildings of the mills have been adapted for agricultural purposes since the decline of the lead industry.

Tan Hill

The old inn at Tan Hill stands approximately 1,738 feet above sea level and looks out across some of the most bleak countryside in Yorkshire. The hill is situated at the top of West Stones Dale and is on the Pennine Way. Tan Hill is the site of an old mine but not a lead mine, although it worked for the lead mining industry. The mine at Tan Hill was a coal mine, its product being used in the smelting of the lead ore extracted in Swaledale and Arkengarthdale. Spoil associated with the old mine is prolific but is mainly grassed over. The Tan Hill coal seam probably belongs to the Arnsbergian stage of the Namurian.

BOWLAND

Clitheroe is famous in geological circles for the reef knolls which are exposed locally and in some cases have been quarried. The knolls are believed to have been formed during the regressive phase at the end of the second cycle of the Dinantian. The knolls

74

begin, apparently, at Clitheroe Castle, which reputedly stands upon one, and continue eastwards occurring at Salt Hill, Warren Hill, Worsaw Hill, Gerna and Twiston. Several of the knolls have been quarried in the past especially around Pimlico.

Limestones of the first cycle of sedimentation which are attributed to the Tournaisian are exposed at the heart of the Clitheroe Anticline near Chatburn. Rocks of the same age have also been identified on the banks of the River Ribble between Gisburn and Sawley.

INGLETON

The small village of Ingleton sits in the shadow of one of Yorkshire's tallest hills, Ingleborough, 2,373 feet. And nearby only a few miles to the north is Yorkshire's tallest hill, Whernside, which stands 2,414 feet high. The village stands on the Craven Fault system. To the south, which is the downthrow side of the fault, the rocks of the Ingleton coalfield formed during Coal Measures Times make up the landscape; while on the north of the fault system the much older Great Scar Limestone, formed during Carboniferous Limestone Times, stands proudly looking out over the village and the younger rocks.

The Ingleton coalfield has not been worked for many years and the rocks which form it are well concealed, except where they are exposed on the flanks of the River Greta well below the village. It is interesting to remember that the Great Scar Limestone which looks down from Ingleborough on to Ingleton also occurs 3,000 feet below the village, thanks to the Craven Fault having preserved the younger rocks on which the village perches.

Thornton Force

The Ingleton coalfield has not been worked for many years its waterfalls, but these same waterfalls are also famous throughout geological circles due to the character of the rock over which the water falls. The controversy which still rages over the age of Yorkshire's oldest rocks is centred around Ingleton and more specifically around the River Greta and the Kingsdale Beck.

Thornton Force is one of a number of waterfalls over which the Kingsdale Beck flows on its way to Ingleton and its junction with the River Greta. The force is the lowest of three falls in a very steep gorge through which the beck has forced its way. By the waterfall it is possible to see clearly the junction between the upturned Ingletonian rocks, which have been tentatively dated as pre-Cambrian, and a conglomerate which is either of Upper Old Red Sandstone Age or, as is most likely, of Carboniferous Age. The conglomerate is succeeded by Carboniferous

Limestone. Further downstream the North Craven Fault crosses the stream and brings limestone and slate adjacent to each other.

The nearby River Greta shares a similarly interesting geology. Again the gorges are cut in the Ingletonian rocks which take the form of grits and slates, and here they are joined by outcrops of the Coniston Limestone which is described as being of Ordovician Age. At one location between Beazley Falls and Ingleton, several Lamprophyre dykes have penetrated the Coniston Limestone and are similar in character to other igneous bodies near Shap in the Lake District.

Both the River Greta and the Kingsdale Beck can be easily reached from the village and can form part of a beautiful and informative walk.

Chapel-le-Dale

Chapel-le-Dale is reached from Ingleton along the B6255, a minor road which runs parallel to the River Greta. The Ingletonian rocks can be seen forming the bed of the river virtually

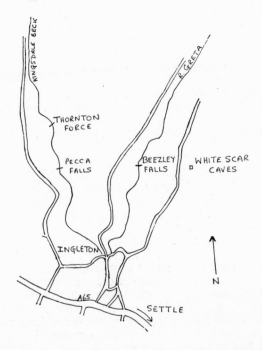

Map showing locations of geological interest in the Ingleton area.

all the way to Chapel-le-Dale.

The Carboniferous Limestone (Great Scar Limestone) is still the dominant feature of the scenery and at the foot of the cliff, known as White Scar, sits White Scar Cave. This cave, which is now open to the public, was first discovered by Christopher Long in August 1923 and although the main features of the cave are set in the limestone, the ancient Ingletonian slate is highlighted in the entrance. Here the roof is formed from the limestone while the slates form the floor and walls. The presence of this cave on the interface between the slate and the limestone bears testimony to the fact that limestone can be easily dissolved in water while slate cannot. White Scar Cave is but one of the many caves in the Great Scar Limestone in the Ingleton area and being a show cave, offers the public an unrivalled opportunity to see the beauty of limestone caves in total safety.

Continuing along the river towards Chapel-le-Dale, the old quarries of the Ingleton 'Granite' Company can be seen by the road. The 'granite' is not a true granite, but is an ancient conglomerate which forms part of the Ingleton complex. Shortly before we reach Chapel-le-Dale we pass the spot where the river appears from beneath the limestone. The river has in fact been running along the surface of the impervious slates underneath the limestone cover from some distance to the north of the village.

Dent Dale

This small but beautiful dale is the home of 'Dent Marble', a black limestone, which was worked at the Marble Mill at Stone House opposite Arten Gill Viaduct on the Settle to Carlisle railway. The marble industry here reached its peak during the middle of the last century.

Ingleborough

Access to the Ingleborough massive is most easily achieved from near the Chapel-le-Dale Road on the outskirts of Ingleton. A track leaves the road near Storrs Cave and proceeds eastwards on to Ingleborough. Storrs Cave is situated in Great Scar Limestone and is sited between the Craven faults. The North Craven Fault is hidden beneath glacial deposits where the track cuts across it towards Crina Bottom.

Above Crina Bottom the Great Scar Limestone gives way to the Yoredale rocks and the succession through sandstone, Main Limestone and gritstone, although the succession is, in places, more imagined than seen. The summit of Ingleborough is in the Lower Howgate Edge Grit. The return trip by the same route is recommended, although an alternative route to Clapham is available.

Norber

Norber is situated approximately 1¼ miles to the north east of Clapham and is reached from Thwaite Lane. This area is carved out of the Great Scar Limestone, but the basement conglomerate underlying the limestone is visible, together with greywackes of Silurian Age, in the lower reaches of Crummack Dale near the village of Wharfe.

The Norber scenery is characterised by the blocks of sandstone which lie upon, and indeed litter the limestone surface. The sandstone is of Silurian Age and is therefore older than the limestone upon which its sits. These blocks were in fact deposited here by the ice sheets of the last glaciation which came to an end around 10,000 years ago. Scratches carved by the ice can often be observed on the limestone, but only where it has been protected from weathering by boulder clay. Many of the sandstone blocks stand upon limestone plinths which protrude from the general surface of the hill. This difference in height is due to the erosion of the limestone surface by the weather since the end of the last glaciation.

NIDDERDALE

Brimham Rocks

Artistic sculpturing by the wind has created this weird collection of rock shapes, which is situated approximately 3½ miles from Pateley Bridge. Access to the rocks can be achieved from either Summer Bridge on the B6165 or from the Pateley Bridge—Ripon road, the B6265.

The rocks are composed of sandstone of the Millstone Grit which has undergone severe erosion by the wind since the end of the last glaciation of the Ice Age. It has been suggested that some of the rocks, which cover over 40 acres, resemble animals rather more intimately than nature intended, thanks to the help of a number of 19th century 'artists'.

Greenhow Hill

This is situated five miles to the west of Pateley Bridge on the road to Grassington. This hill, together with other sites on Grassington Moor, formed an important part of the lead mining industry more than a century ago. Spoil heaps associated with this long departed industry may still be seen. Stump Cross Cavern on Greenhow Hill is yet another example of a limestone cavern once considered to be the most spectacular in Yorkshire.

Knaresborough

A very beautiful gateway to the Yorkshire Dales, Knaresborough stands squarely upon the banks of the River Nidd.

Geologically, it is based upon the Magnesian Limestone which belongs to the Permian Period of geological history. The relationship between the Magnesian Limestone and the underlying Carboniferous rocks is highlighted in the cliff below Knaresborough Castle. Here the limestone can be observed overlying the sandstone of the Millstone Grit. The sandstone is believed to be of the Middle Grits and the limestone rests unconformably upon it. The red colour frequently found in rocks which lie immediately below Permian rocks is also visible.

Lofthouse—Pateley Bridge
The road between Pateley Bridge and Lofthouse runs alongside the River Nidd and highlights the rocks of the Millstone Grit. As with Greenhow Hill, the upper reaches of Nidderdale have been the source of much lead ore which was worked extensively in the past.

To the north of Lofthouse the Millstone Grit continues to predominate, but Carboniferous Limestones from near the top of the succession can be observed beneath the sandstone cover, where streams have carved their path off the hillside. Both the Goyden Pot and the Manchester Hole, which are yet further examples of limestone caverns, occur towards the northern end of the valley. Numerous examples of corals have been found in the Carboniferous Limestone in the river banks near the Goyden Pot.

RIBBLESDALE

Horton-in-Ribblesdale
Rocks of Silurian Age known as the Horton Flags, are exposed to the north of Little Stainforth between Settle and Horton-in-Ribblesdale. These slates have been worked quite extensively in a number of very old quarries along the west bank of the River Ribble.

Arcow Wood Quarry is famous for the relationship it has illustrated between the Silurian rocks and the Carboniferous Limestone which overlies them unconformably. However, the same relationship exists between these rocks for some distance to the north, but is largely unseen. The Silurian rocks are extremely folded and contorted which contrasts with the horizontal bedding of the limestone.

Before reaching Horton, the Silurian rocks give way to the Coniston Limestone, and the Ingletonian rocks previously described.

SEDBERGH

Sedbergh is an attractive market town approximately twenty miles north of Ingleton on the A683. The town is famous for its school which was founded in 1525. One of its pupils, Adam Sedgwick, who was born in nearby Dent subsequently became a famous and respected geologist, during the early years of the last century. This town is situated close to some of the most significant exposures of the pre-Carboniferous rocks in the Pennines. Unfortunately Sedbergh, due to the reorganisation of the county boundaries, now forms part of Cumbria. It seems strange that the politicians or civil servant's pen can at a stroke achieve what thousands of years of history have failed to do.

Backside and Blue Caster

The Backside Beck is reached from Sedbergh by leaving the town and travelling north eastwards along the A683. The beck joins the river Rawthey approximately $\frac{1}{4}$ mile beyond the 'Cross Keys', having flowed between Yarlside and Wandale Hill. The beck flows over rocks which have been described as mudstones of the Caradoc and Ashgill Series of the Ordovician, while the adjacent Wandale Hill is hewn out of shales and grits of the Llandovery, Wenlock and Ludlow Series of the Silurian. Igneous rocks of the Cautley Volcanic Group cut across the upper reaches of the beck and are preceded by several minor intrusions of felsite.

On returning downhill to the main road, it is possible to ascend the hill on the east side of the road opposite Backside Beck, this is called Blue Caster. The lower slopes of the hill are in Silurian slates and shales, but the upper and indeed the major area of the hill is formed from an intrusion of dolerite.

Silurian rocks and basement conglomerates of the Carboniferous are also exposed in a number of stream beds to the east of Sedbergh, adjacent to the A684. Hebblethwaite Hall Gill offers exposures of the conglomerate, and Hole Beck flows through Silurian sediments.

SWALEDALE

Swaledale is arguably the most beautiful of Yorkshire's major dales. Unlike Wensleydale and to a lesser degree Teesdale, Swaledale is a narrow valley and its beauty is primarily due to the ice sculptured scars which overlook it. The rocks exposed in Swaledale belong mainly to the Yoredale Series which is no surprise bearing in mind its northerly position. Although the dale has been cut deep, at no point is rock lower than the Great Scar

Limestone exposed and this is not extensively exposed at any point. The dale is flanked on either side by the moorlands which are based upon the rocks of the Millstone Grit. In the depths of the dale, boulder clay overlies the bed rock, hiding it from view.

Gunnerside

Gunnerside is one of the most beautiful spots in Swaledale and is a large village which was developed during the lead mining days of the 18th and 19th centuries. Visual evidence of the lead mining industry is still plentiful and can be seen in the form of spoil heaps which stand out on the high ground alongside Gunnerside Gill.

From Gunnerside it is a short distance along the B6270 to Oxnop Gill. By walking up the Gill it is possible to follow part of the lower Yoredale succession. This begins with the Hardraw Scar Limestone, which is followed here by the Simonstone Limestone and the Middle Limestone. The lower series of waterfalls is on the Simonstone Limestone, the upper series being on the Middle Limestone. The Great Scar Limestone is found forming the bed of the stream at its junction with the River Swale.

TEESDALE

Teesdale has never been more than half a Yorkshire Dale because the River Tees has for centuries marked the boundary between Yorkshire and Durham. Recent boundary changes have resulted in the whole of Teesdale becoming part of Durham; however, the interesting geological features which occur in the dale are sufficiently special to warrant inclusion here.

Eggleston

The River Tees has cut its path through sandstone of the Yoredale Series and this can be seen at the bridge where the B6281 crosses the river. The Eggleston Burn, which flows into the Tees from the north, contains exposures in rocks of the upper succession of the Yoredale Series, and also highlights the Cleveland Dyke which cuts across the stream approximately one mile north of the village. Another dyke is said to cross the burn further to the north.

Middleton-in-Teesdale—High Force

The village of Middleton-in-Teesdale grew under the influence of the lead mining industry, and was the principal local headquarters for the London Lead Company which operated for almost 300 years. The remains of this company's activities may be seen by taking the road which follows the Hudeshope Beck. Spoil heaps at the head of the beck can be examined and specimens of the prominent vein minerals may be collected.

The Whin Sill, an igneous intrusion which has been dated as late Carboniferous to early Permian Age, forms a number of crags and scars to the south of Middleton which run parallel with the River Tees. At High Force, the magnificent waterfall which is four miles to the west of Middleton, the sill crosses the river and can be examined. It is reached from the B6277. The sill has been quarried locally for road stone.

WENSLEYDALE

Wensleydale is the largest of the dales and offers a great variety of scenery. The lower eastern region of the dale is wide and suggests little of the wild beauty it manages to achieve near its head away to the west. The dale is cut through rocks of the Yoredale Series, the Great Scar Limestone only being exposed towards the head of the dale where it has been deeply incised. The weathering characteristics of the rocks formed under the cyclic conditions can be clearly seen on the hillsides. Projections of the hard sandstones or limestones are separated by recessed bands which appear characterless. In fact the recessed bands represent the shales and mudstones, which being softer have weathered more readily than their harder sandstone and limestone relatives. The result of the cyclic pattern of sedimentation is therefore preserved here for everyone to see.

Waterfalls are an integral part of the Wensleydale scene and as always offer the geologist the opportunity to examine the bed rocks. At Aysgarth Force on the River Ure the water falls over limestone which belongs to the upper beds of the Great Scar Limestone. At Hardraw Force, to the north of Hawes, the rocks exposed are of the Yoredale Series and indeed the Hardraw Scar Limestone takes its name from here. Similar exposures occur on the Whity Gill which rises on Abbotside Common above Askrigg. The basic succession ranges from the Great Scar Limestone at the base of the hill, through sandstone, shale, Hardraw Limestone, sandstone, Simonside Limestone and sandstone. The Hardraw Limestone is also exposed by the waterfall up Cotterdale, to the north west of Hawes.

Further details of the local geology including glacial features of the terrain can be obtained from the Yorkshire Dales National Park Centre at Aysgarth.

WHARFEDALE

Upper Wharfedale is a typical limestone dale. The Great Scar Limestone is well represented and is overlain by rocks of the Yoredale Series, while rocks of the Millstone Grit Series top some of the surrounding moorland.

Grassington

Grassington is a village which thrived during the days of the lead mining industry although this association is not obvious today unless you take the minor road to the north east out of the village onto Grassington Moor. Old mine workings are abundant here, and as always on mining terrain, beware of old unguarded shafts which may be several hundred feet deep. Lead and copper minerals may still be located on the old mine dumps, together with fluorite, calcite and some barite. Aurichalcite, a copper mineral, is just one example of the less spectacular, but more unusual minerals which occur here. This mineral has been located near Yarnbury.

Kilnsey Crag

Kilnsey Crag stands on the west bank of the River Wharfe between Grassington and Kettlewell. The crag is hewn out of the Great Scar Limestone and has been well rounded by glaciation. A nearby stream contains many boulders of Silurian rocks, which tends to suggest that the upper surface of the Silurian rocks underlying the limestone is not far beneath the bed of the stream, and hence not far below the lowest exposed limestone of the nearby crag.

Littondale

The River Skirfare is a tributary of the Wharfe and flows through Littondale, before joining the river near Conistone. This is a typical limestone dale and has many caverns and waterfalls.

From Arncliffe, excellent exposures of the Great Scar Limestone, Yoredale Series and Millstone Grit Series can be seen by following the footpath above Cowside Beck from Arncliffe. Some old depressions on nearby Fountains Fell represent places where small coal seams in the Millstone Grit Series were once worked.

Ilkley

From Bolton Abbey eastwards the River Wharfe leaves the area of limestone dominance, and flows through an area where the scenery is made up of rocks of the Millstone Grit Series. The Rough Rock and Middle Grits of this Series, interbedded with shale, form the landscape of Rombolds Moor south of Ilkley.

Glossary of Terms

Arenaceous Rocks: A number of detrial sedimentary rocks, normally sandstones, composed of particles which range in size from 0.07 mm to 2 mm. These rocks are accumulated by wind or water action.

Argillaceous Rocks: These are detrial sedimentary rocks with particle sizes from 0.07 mm to less than 0.004 mm. Rocks of this grade include clays, mudstones, shales etc.

Barite: Barium Sulphate Ba SO₄. A common mineral in sulphide ore veins.

Bedding Plane: A bedding plane is a surface which lies parallel to the surface of deposition. The plane may or may not be visible, but often marks a change in the character of the sedimentation.

Boulder Bed: A sedimentary rock consisting of large pieces of rock fragments above 256 mm in size, set in a fine matrix.

Breccia: Any sedimentary rock composed of angular rock fragments cemented into a fine matrix.

Calcite: Calcite is the main constituent in limestone, but also occurs in mineral veins and as stalactites and stalagmites. Chemically it is Calcium Carbonate, CaCo₃.

Chert: A cryptocrystalline form of silica which may have inorganic or organic origins. It usually occurs as bands or nodules in sedimentary rocks.

Conglomerate: Any sedimentary rock composed of well rounded pebbles cemented into a fine matrix.

Cyclic sedimentation: This term is used to describe a sequence of deposition of sediment which progresses from one extreme type to another, and then returns to the first type again. This may be visualised as follows:-
  1. Mudstone
  2. Siltstone
  3. Sandstone
  4. Coal
  3. Sandstone
  2. Siltstone
  1. Mudstone

Delta: A sedimentary deposit formed at the mouth of a river.

Erosion: This is defined as the erosion of a land surface by the action of transported debris. It is therefore particles of weathered material which do the damage, when carried by water and wind.

Fault: A fracture in the rocks of the earth's crust along which there has been a visible amount of displacement.

Felsite: A fine grained acid or intermediate igneous rock, normally found in the form of a dyke.

Fire Clay: This is a fossil soil usually associated with coal seams.

Fluorite: Calcium Fluorite Ca F_2 is a common mineral in ore veins.

Galena: Lead Sulphide Pb S. This is the prime ore of lead and has been worked extensively in the Pennines.

Gangue: The part of a mineral vein or deposit which is not the prime purpose of the extraction, in other words the unwanted mineral or rock.

Ganister: A seatearth found below coal seams which usually consists of pure silica sand.

Grit: A word applied to arenaceous rocks which are composed of angular grains.

Ice Age: A time when glacial ice reached out to cover areas not normally ice covered.

Inlier: An exposure of older rocks completely surrounded by younger rocks.

Lamprophyres: An igneous rock, medium grained and usually containing phenocrysts of feldspar. This rock is often found in a highly decomposed state.

Marble: A name given to a limestone which has been metamorphosed. Unfortunately many limestones which have not been under the influence of metamorphism, but take a polish and have been used for decorative purposes, have also been called marble (Dent Marble).

Ore: The name given collectively to the various constituents which make up an economical mineral vein.

Orogeny: This term relates to every aspect of a mountain building period.

Outlier: A small area of younger rocks completely surrounded by older rocks.

Peneplain: The name given to a land mass which has been eroded virtually to sea level, with minimal surface relief.

Tufa: A name given to deposits of Calcium Carbonate when formed from Calcium Bicarbonate.

Unconformity: Simply defined, an unconformity is evidence that between the rocks below and above it, there was a period of time not accounted for in the succession. Hence two rocks which now lie one above the other, may have their origins millions of years apart.

Bibliography and Further Reading

Dunham, K. C., Hemingway, J. E., Versey, H. C. and Wilcockson, W. H. 1953. *A guide to the geology of the district around Ingleborough.* Proc. Yorks. geol. Soc. 29, 77-115.

Edwards, M. A. and Trotter, F. M. 1954. *British Regional Geology, The Pennines and Adjacent Areas.* H.M.S.O., London.

Kendal, P. F and Wroot, H. E. 1924. *The Geology of Yorkshire.* Vol. 2. First published privately by the authors. Republished 1972 by E. P. Publishing Ltd., Wakefield, Yorkshire.

King, W. B. R., and Wilcockson, W. H. 1934. *The Lower Palaeozoic rocks of Austwick and Horton-in-Ribblesdale, Yorkshire.* Q. Jl geol. Soc. Lond. 90, 7-30.

Mitchell, G. H. 1967. *The Caledonian orogeny in Northern England.* Proc. Yorks. geol. Soc. 36, 135-8.

Muir Wood, R. and Rodgers, P. R. 1978. *On the Rocks* (A Geology of Britain) B.B.C. Publications, London.

Oldham, K. 1972. *The Pennine Way,* Dalesman, Clapham, Yorkshire.

Rayner, D. H. 1953. *The Lower Carboniferous rocks in the north of England: a review.* Proc. Yorks. geol. Soc. 28, 231-315.

Rayner, D. H. and Hemingway, J. E., (editors) 1974. *The Geology and Mineral Resources of Yorkshire.* Yorkshire Geological Society.

Rodgers, P. R. 1976. *Derbyshire Geology.* Dalesman, Yorkshire.

Rodgers, P. R. 1975. *Yorkshire Minerals.* Dalesman, Yorkshire.

Stamp, L. Dudley, 1946. *Britain's Structure and Scenery.* Fontana, London.

Wills, L. J. 1951. *A palaeogeographical atlas of the British Isles and adjacent parts of Europe.* Blackie, London.

Index